Temperature Measurement

Temperature Measurement

Béla G. Lipták
EDITOR-IN-CHIEF

CHILTON BOOK COMPANY | RADNOR, PENNSYLVANIA

To my colleagues, the process control and instrument engineers, with the hope that they will make the world a better, safer and happier place by applying the knowledge found on these pages

Published in Radnor, Pennsylvania 19089, by Chilton Book Company

Designed by Arlene Putterman
Manufactured in the United States of America

Library of Congress Cataloging-in-Publication Data
Temperature measurement / Béla G. Lipták, editor-in-chief.
p. cm.
"An offprint book from Instrument engineers' handbook, third ed."—CIP data sheet.
Includes index.
ISBN 0-8019-8385-1 (pbk.)
1. Process control—Equipment and supplies. 2. Temperature measuring instruments. I. Lipták, Béla G. II. Title: Temperature measurement.
TS156.8.T42 1993
536'.5'0287—dc20 92-56586
CIP

1 2 3 4 5 6 7 8 9 0 2 1 0 9 8 7 6 5 4 3

CONTENTS

CONTRIBUTORS

THOMAS J. CLAGGETT BSEE, Application Specialist, Honeywell Inc.

WILSON A. CLAYTON BSChE, MSME, Chief Engineer, Hy-Cal Engineering

H. L. DANEMAN BChE, PE, Principal, LabPlan

BÉLA G. LIPTÁK ME, MME, PE, President, Lipták Associates P. C.

DIETER RALL BSME, MSME, PE, General Manager, Trans-Met Engineering Inc.

ROBERT W. WORRALL BA, PE, Principal Instrument Engineer, Catalytic Inc.

INTRODUCTION

This is the first ''spinoff'' book from the *Process Measurement and Analysis* volume of the *Instrument Engineers' Handbook*. Its purpose is to provide timely and cost-effective information on selected subject matters. The *Instrument Engineers' Handbook* was first published in 1968–69, with a supplemental volume issued in 1972; then it was completely revised in 1982–85. The third complete revision is scheduled for 1994–95. While a complete handbook cannot be updated more often than once a decade, some of its chapters can. The chapters covering topics that are fast changing, and should therefore be updated more often, were selected as ''spinoff'' volumes.

Another consideration in selecting chapters for separate publication was economic: If a reader works with only one category of instruments on a daily basis, it is probably efficient to visit the company library when information is needed on other categories of devices. But the chapter dealing with the area of his or her specific interest, should be permanently on that reader's desk. Publishing ''spinoff'' volumes provides readers with this flexibility at low cost.

In selecting subjects for separate publication as ''spinoff'' volumes, there was little question that temperature measurement should be one of them. This is partly because temperature is one of the most often used sensors, partly because it provides critical measurements for chemical reactors, distillation and other important processes, but most importantly, because it is a fast-changing technology. The addition of the microprocessor has changed—and is still changing—the capabilities of the temperature transmitters; the time is near when each smart temperature transmitter will be directly connected through any data highway to any DCS system. The smart transmitters will improve accuracy not only because they eliminate the need for A/D conversion, but also because they will be able to self-calibrate their spans, self-diagnose their sensor integrity and even make automatic selections among their multiple sensors, to provide automatic replacement or to match the temperature conditions occurring in the process. The end result will be a combination of reduced measurement error and higher rangeability. Smart transmitters can also improve the lead wire compensation in RTDs (3 and 4 wires) or detect thermocouple burn-out.

Other changes that are expected in the field of temperature measurement include the more extensive use of fiber optics, both as a means of signal transmission and as a means of detecting IR radiation from hard-to-reach areas. Other rapidly changing areas include the use of integrated circuitry sensors which, due to their low unit costs and strong, linear signals are ideally suited not only as cold junction compensators, but also as direct thermometers. The combination of battery-operated microprocessor memory and small thermistor sensors can be packaged to travel with the process materials in batch processes to memorize their temperature history. Further progress is expected in the high-precision detection of temperature differentials, which is needed in all heat-load, cooling-load, or exothermic heat release calculations. Similarly, there will be progress in detecting the temperature of moving surfaces, either by non-contacting sensors or by radio transmission of the readings from transmitters mounted on the moving equipment. My hope is that we will be able to report on these and also on the developments of the international marketplace at frequent intervals.

Béla Lipták
Stamford, Connecticut, May, 1993

Temperature Measurement

TABLE A

Orientation Table for Temperature Sensors

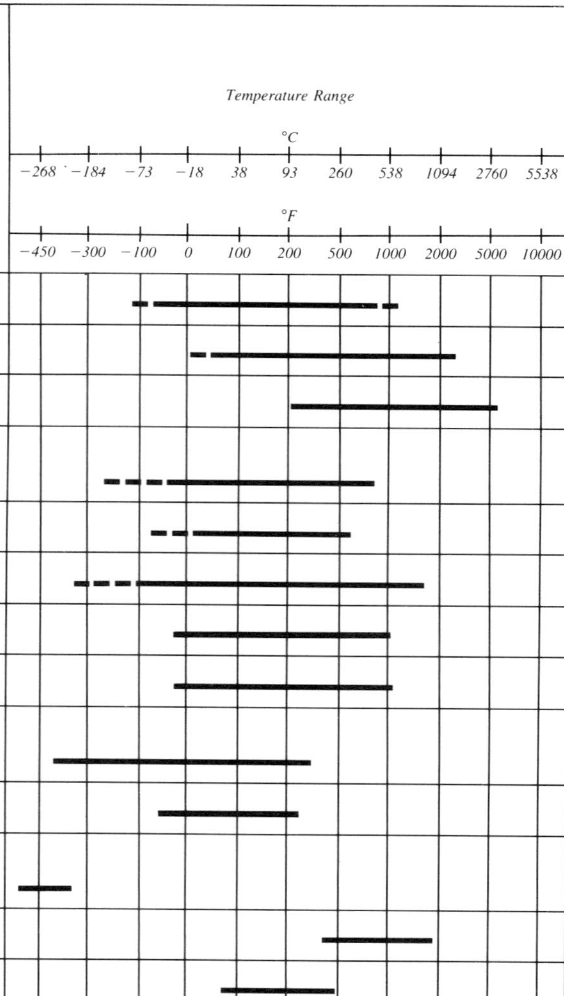

Type	Available Span: Under 100°F (38°C)	Available Span: Between 100 and 1000°F	Available Span: Above 1000°F (538°C)	Accuracy: % Full Scale or % of Span	Accuracy: Best Attainable in °F (°C = 5/9°F)	Cost ($): Under $200	Cost ($): Between $200 and $1000	Cost ($): Above $1000	Sensor Size: Small	Sensor Size: Medium	Sensor Size: Large	Available With: Recorder	Available With: Indicator	Available With: Controller	Stability	Repeatability	Response Time	Sensitivity	① Interchangeable	Linear	② Complete System	③ Maximum Distance to Readout in Feet (0.305m)
Bimetallic Elements	√	√		1–2	1	√			√	√		√	√	√	E	F	G	G		N	√	N
Color Indicators	√				8	√							√					F			√	N
Fiber-Optic		√	√	1	0.2			√	√			√	√	√	F	F	E	G		N	√	1000+
Filled Elements Liquid	√	√		0.5–2	0.1	√	√		√	√		√	√	√	E	F	F	G		√	√	40
Vapor	√	√		0.5–2	0.6	√	√			√		√	√	√	E	F	G	F		N	√	200
Gas		√	√	0.5–2	1.2	√	√				√	√	√	√	E	F	G	F		√	√	150
Mercury	√	√	√	0.5–2	0.25	√	√		√	√		√	√	√	E	F	F	G		√	√	50
Glass-Stem Therm.	√	√		0.1–2	0.01	√	√			√			√		E	E	G	G	√	√	√	N
Integrated Circuits Diodes	√	√		0.2–2	0.2		√		√				√		G	G	G	E		√	N	
Transistors	√	√		2	1		√		√				√		F	G	G	E		√	N	
Misc.— Carbon Resistors	√				0.5		√		√	√			√		F,G	G	G	G		N	N	
Fluidic Sensor		√		2			√				√		√		F	F	F	F		N	N	
Liquid Crystals	√				1	√					√		√		E	G,E	G	E,G			√	

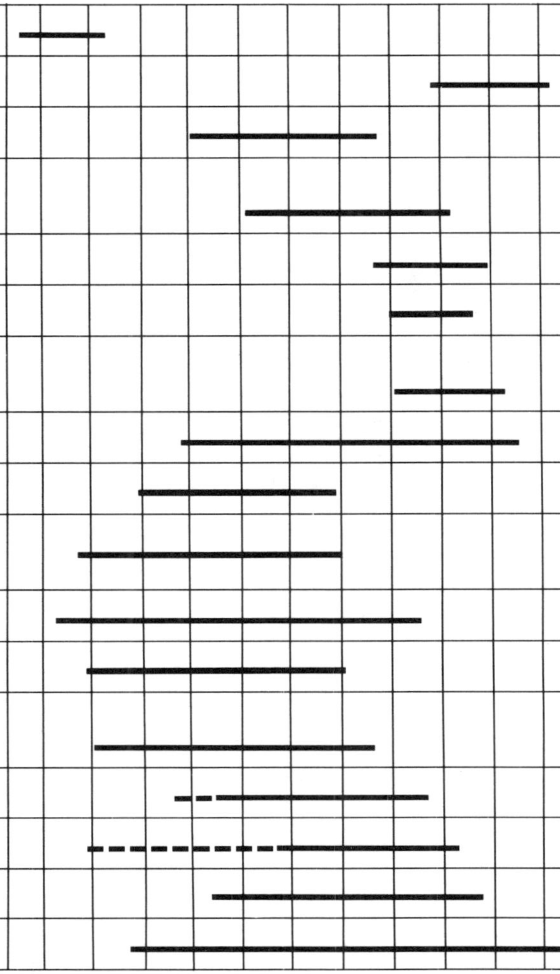

Paramagnetic Salts	√			1	0.005			√			√		√		G	G,E	E	F,G		N	√	
Spectroscopy			√	1				√			√		√		G	G,E	G			N	√	
Thermography		√			1	√	√						√					F			√	
Pyrometers Suction		√	√	2	20			√			√	√	√		F	G	F	P		N	N	
Pneumatic		√	√	2	25			√			√		√		F	G	F	P		N	N	
Pyrometric Cones	√	√			5	√			√	√			√		E	F	F	F			√	N
Radiation Pyrometers Optical & Ratio			√	1–2	2–10			√			√	√	√	√	F	F	G	G		N	√	100
Narrow & Wide Band		√	√	0.5–2	5			√			√	√	√	√	F	F	E	G		N	√	100
Quartz Crystals	√	√		0.1	0.2			√		√	√	√	√		E	G	G	E	√	√	N	1000
Resistance Bulbs Nickel	√	√		0.25	0.3	√	√	√		√		√	√	√	G,E	E	G	E	√	N	N	1000
Platinum	√	√	√	0.15	0.2		√	√		√		√	√	√	E	E	G	G,E	√	√	N	3000
Thermistors	√	√		0.2	0.02	√	√		√			√	√	√	F	G	E	E		N	N	3000
Thermocouples Type T	√	√		0.1	1.5	√	√		√	√		√	√	√	G	G	G	G	√	N	N	3000
Type J	√	√	√	0.1	2.5	√	√		√	√		√	√	√	G	G	G	G	√	N	N	3000
Type K	√	√	√	0.1	2.5		√		√	√		√	√	√	G	G	G	G	√	√	N	3000
Types R & S		√	√	0.1	4		√		√	√		√	√	√	G	G	E	E	√	N	N	3000
Ultrasonic	√	√	√	5				√			√		√		G	F,G	E	E,G		N	N	

Terminology

N—No or None
E—Excellent
G—Good
F—Fair

① Interchangeable sensor, without recalibration of entire system.
② System is complete when sensor and readout is sold as a single unit. When several readouts can be used with the same sensor, system is not considered to be complete.
③ Without special compensation.

Solid line: Recommended
Dashed line: Available but not recommended

TABLE B

Temperature Conversion Table

When converting any temperature, find the the boldface value of the temperature to be converted and look to the left for its °C equivalent or to the right for its °F equivalent. Temperatures not listed can be converted using °F = (9°C/5) + 32 or °C = 5(°F-32)/9.

C		F	C		F	C		F	C		F	C		F	C		F	C		F	C		F	C		F
−273.1	**−459.4**		−17.8	**0**	32	10.0	**50**	122.0	38	**100**	212	260	**500**	932	538	**1000**	1832	816	**1500**	2732	1093	**2000**	3632	1371	**2500**	4532
−268	**−450**		−17.2	**1**	33.8	10.6	**51**	123.8	43	**110**	230	266	**510**	950	543	**1010**	1850	821	**1510**	2750	1099	**2010**	3650	1377	**2510**	4550
−262	**−440**		−16.7	**2**	35.6	11.1	**52**	125.6	49	**120**	248	271	**520**	968	549	**1020**	1868	827	**1520**	2768	1104	**2020**	3668	1382	**2520**	4568
−257	**−430**		−16.1	**3**	37.4	11.7	**53**	127.4	54	**130**	266	277	**530**	986	554	**1030**	1886	832	**1530**	2786	1110	**2030**	3686	1388	**2530**	4586
−251	**−420**		−15.6	**4**	39.2	12.2	**54**	129.2	60	**140**	284	282	**540**	1004	560	**1040**	1904	838	**1540**	2804	1116	**2040**	3704	1393	**2540**	4604
−246	**−410**		−15.0	**5**	41.0	12.8	**55**	131.0	66	**150**	302	288	**550**	1022	566	**1050**	1922	843	**1550**	2822	1121	**2050**	3722	1399	**2550**	4622
−240	**−400**		−14.4	**6**	42.8	13.3	**56**	132.8	71	**160**	320	293	**560**	1040	571	**1060**	1940	849	**1560**	2840	1127	**2060**	3740	1404	**2560**	4640
−234	**−390**		−13.9	**7**	44.6	13.9	**57**	134.6	77	**170**	338	299	**570**	1058	577	**1070**	1958	854	**1570**	2858	1132	**2070**	3758	1410	**2570**	4658
−229	**−380**		−13.3	**8**	46.4	14.4	**58**	136.4	82	**180**	356	304	**580**	1076	582	**1080**	1976	860	**1580**	2876	1138	**2080**	3776	1416	**2580**	4676
−223	**−370**		−12.8	**9**	48.2	15.0	**59**	138.2	88	**190**	374	310	**590**	1094	588	**1090**	1994	886	**1590**	2894	1143	**2090**	3794	1421	**2590**	4694
−218	**−360**		−12.2	**10**	50.0	15.6	**60**	140.0	93	**200**	392	316	**600**	1112	593	**1100**	2012	871	**1600**	2912	1149	**2100**	3812	1427	**2600**	4712
−212	**−350**		−11.7	**11**	51.8	16.1	**61**	141.8	99	**210**	410	321	**610**	1130	599	**1110**	2030	877	**1610**	2930	1154	**2110**	3830	1432	**2610**	4730
−207	**−340**		−11.1	**12**	53.6	16.7	**62**	143.6	100	**212**	413	327	**620**	1148	604	**1120**	2048	882	**1620**	2948	1160	**2120**	3848	1438	**2620**	4748
−201	**−330**		−10.6	**13**	55.4	17.2	**63**	145.4	104	**220**	428	332	**630**	1166	610	**1130**	2066	888	**1630**	2966	1166	**2130**	3866	1443	**2630**	4766
−196	**−320**		−10.0	**14**	57.2	17.8	**64**	147.2	110	**230**	446	338	**640**	1184	616	**1140**	2084	893	**1640**	2984	1171	**2140**	3884	1449	**2640**	4784
−190	**−310**		− 9.44	**15**	59.0	18.3	**65**	149.0	116	**240**	464	343	**650**	1202	621	**1150**	2102	899	**1650**	3002	1177	**2150**	3902	1454	**2650**	4802
−184	**−300**		− 8.89	**16**	60.8	18.9	**66**	150.8	121	**250**	482	349	**660**	1220	627	**1160**	2120	904	**1660**	3020	1182	**2160**	3920	1460	**2660**	4820
−179	**−290**		− 8.33	**17**	62.6	19.4	**67**	152.6	127	**260**	500	354	**670**	1238	632	**1170**	2138	910	**1670**	3038	1188	**2170**	3938	1466	**2670**	4838
−173	**−280**		− 7.78	**18**	64.4	20.0	**68**	154.4	132	**270**	518	360	**680**	1256	638	**1180**	2156	916	**1680**	3056	1193	**2180**	3956	1471	**2680**	4856
−169	**−273**	−459.4	− 7.22	**19**	66.2	20.6	**69**	156.2	138	**280**	536	366	**690**	1274	643	**1190**	2174	921	**1690**	3074	1199	**2190**	3974	1477	**2690**	4874
−168	**−270**	−454	− 6.67	**20**	68.0	21.1	**70**	158.0	143	**290**	554	371	**700**	1292	649	**1200**	2192	927	**1700**	3092	1204	**2200**	3992	1482	**2700**	4892
−162	**−260**	−436	− 6.11	**21**	69.8	21.7	**71**	159.8	149	**300**	572	377	**710**	1310	654	**1210**	2210	932	**1710**	3110	1210	**2210**	4010	1488	**2710**	4910

−157	**−250**	−418	− 5.56	**22**	71.6	22.2	**72**	161.6	154	**310**	590	382	**720**	1328	660	**1220**	2228	938	**1720**	3128	1216	**2220**	4028	1493	**2720**	4928
−151	**−240**	−400	− 5.00	**23**	73.4	22.8	**73**	163.4	160	**320**	608	388	**730**	1346	666	**1230**	2246	943	**1730**	3146	1221	**2230**	4046	1499	**2730**	4946
−146	**−230**	−382	− 4.44	**24**	75.2	23.3	**74**	165.2	166	**330**	626	393	**740**	1364	671	**1240**	2264	949	**1740**	3164	1227	**2240**	4064	1504	**2740**	4964
−140	**−220**	−364	− 3.89	**25**	77.0	23.9	**75**	167.0	171	**340**	644	399	**750**	1382	677	**1250**	2282	954	**1750**	3182	1232	**2250**	4082	1510	**2750**	4982
−134	**−210**	−346	− 3.33	**26**	78.8	24.4	**76**	168.8	177	**350**	662	404	**760**	1400	682	**1260**	2300	960	**1760**	3200	1238	**2260**	4100	1516	**2760**	5000
−129	**−200**	−328	− 2.78	**27**	80.6	25.0	**77**	170.6	182	**360**	680	410	**770**	1418	688	**1270**	2318	966	**1770**	3218	1243	**2270**	4118	1521	**2770**	5018
−123	**−190**	−310	− 2.22	**28**	82.4	25.6	**78**	172.4	188	**370**	698	416	**780**	1436	693	**1280**	2336	971	**1780**	3236	1249	**2280**	4136	1527	**2780**	5036
−118	**−180**	−292	− 1.67	**29**	84.2	26.1	**79**	174.2	193	**380**	716	421	**790**	1454	699	**1290**	2354	977	**1790**	3254	1254	**2290**	4154	1532	**2790**	5054
−112	**−170**	−274	− 1.11	**30**	86.0	26.7	**80**	176.0	199	**390**	734	427	**800**	1472	704	**1300**	2372	982	**1800**	3272	1260	**2300**	4172	1538	**2800**	5072
−107	**−160**	−256	− 0.56	**31**	87.8	27.2	**81**	177.8	204	**400**	752	432	**810**	1490	710	**1310**	2390	988	**1810**	3290	1266	**2310**	4190	1543	**2810**	5090
−101	**−150**	−238	0	**32**	89.6	27.8	**82**	179.6	210	**410**	770	438	**820**	1508	716	**1320**	2408	993	**1820**	3308	1271	**2320**	4208	1549	**2820**	5108
− 95.6	**−140**	−220	0.56	**33**	91.4	28.3	**83**	181.4	216	**420**	788	443	**830**	1526	721	**1330**	2426	999	**1830**	3326	1277	**2330**	4226	1554	**2830**	5126
− 90.0	**−130**	−202	1.11	**34**	93.2	28.9	**84**	183.2	221	**430**	806	449	**840**	1544	727	**1340**	2444	1004	**1840**	3344	1282	**2340**	4244	1560	**2840**	5144
− 84.4	**−120**	−184	1.67	**35**	95.0	29.4	**85**	185.0	227	**440**	824	454	**850**	1562	732	**1350**	2462	1010	**1850**	3362	1288	**2350**	4262	1566	**2850**	5162
− 78.9	**−110**	−166	2.22	**36**	96.8	30.0	**86**	186.8	232	**450**	842	460	**860**	1580	738	**1360**	2480	1016	**1860**	3380	1293	**2360**	4280	1571	**2860**	5180
− 73.3	**−100**	−148	2.78	**37**	98.6	30.6	**87**	188.6	238	**460**	860	466	**870**	1598	743	**1370**	2498	1021	**1870**	3398	1299	**2370**	4298	1577	**2870**	5198
− 67.8	**− 90**	−130	3.33	**38**	100.4	31.1	**88**	190.4	243	**470**	878	471	**880**	1616	749	**1380**	2516	1027	**1880**	3416	1304	**2380**	4316	1582	**2880**	5216
− 62.2	**− 80**	−112	3.89	**39**	102.2	31.7	**89**	192.2	249	**480**	896	477	**890**	1634	754	**1390**	2534	1032	**1890**	3434	1310	**2390**	4334	1588	**2890**	5234
− 56.7	**− 70**	− 94	4.44	**40**	104.0	32.2	**90**	194.0	254	**490**	914	482	**900**	1652	760	**1400**	2552	1038	**1900**	3452	1316	**2400**	4352	1593	**2900**	5252
− 51.1	**− 60**	− 76	5.00	**41**	105.8	32.8	**91**	195.8				488	**910**	1670	766	**1410**	2570	1043	**1910**	3470	1321	**2410**	4370	1599	**2910**	5270
− 45.6	**− 50**	− 58	5.56	**42**	107.6	33.3	**92**	197.6				493	**920**	1688	771	**1420**	2588	1049	**1920**	3488	1327	**2420**	4388	1604	**2920**	5288
− 40.0	**− 40**	− 40	6.11	**43**	109.4	33.9	**93**	199.4				499	**930**	1706	777	**1430**	2606	1054	**1930**	3506	1332	**2430**	4406	1610	**2930**	5306
− 34.4	**− 30**	− 22	6.67	**44**	111.2	34.4	**94**	201.2				504	**940**	1724	782	**1440**	2624	1060	**1940**	3524	1338	**2440**	4424	1616	**2940**	5324
− 28.9	**− 20**	− 4	7.22	**45**	113.0	35.0	**95**	203.0				510	**950**	1742	788	**1450**	2642	1066	**1950**	3542	1343	**2450**	4442	1621	**2950**	5342
− 23.3	**− 10**	14	7.78	**46**	114.8	35.6	**96**	204.8				516	**960**	1760	793	**1460**	2660	1071	**1960**	3560	1349	**2460**	4460	1627	**2960**	5360
− 17.8	**0**	32	8.33	**47**	116.6	36.1	**97**	206.6				521	**970**	1778	799	**1470**	2678	1077	**1970**	3578	1354	**2470**	4478	1632	**2970**	5378
			8.89	**48**	118.4	36.7	**98**	208.4				527	**980**	1796	804	**1480**	2696	1082	**1980**	3596	1360	**2480**	4496	1638	**2980**	5396
			9.44	**49**	120.2	37.2	**99**	210.2				532	**990**	1814	810	**1490**	2714	1088	**1990**	3614	1366	**2490**	4514	1643	**2990**	5414

1 Application and Selection

D. RALL (1969) **B. G. LIPTÁK** (1982, 1993)

Temperature is an expression denoting a physical condition of matter, just as are mass, dimensions, and time. Yet, the idea of temperature is a relative one, arrived at by a number of conflicting theories. The classic theory depicts heat as a form of energy associated with the activity of the molecules of a substance. These minute particles of all matter are assumed to be in continuous motion which is sensed as heat. Temperature is a measure of this heat.

To standardize on the temperature of objects under varying conditions, several scales have been devised. The Fahrenheit scale arbitrarily assigns the number 32 to the freezing point of water and the number 212 to the boiling point of water and divides the interval into 180 equal parts. The Centigrade or Celsius scale calls the freezing point of water 0 and its boiling point 100.

In line with the classic theory, some relation to the point where molecular motion is at a minimum had to be established, and the Kelvin scale, using Centigrade divisions, was drawn. Zero Kelvin was determined to be −273.19°C. The Rankine scale places its zero at −459.61°F using Fahrenheit divisions in the same arbitrary way in which Lord Kelvin used the Celsius scale.

THE INTERNATIONAL PRACTICAL TEMPERATURE SCALE (IPTS)

The International Practical Temperature Scale is the basis of most present-day temperature measurements. The scale was established by an international commission in 1948 with a text revision in 1960.[1] A revision[2] of the scale was formally adopted in 1990 and is reproduced in Tables 1a and 3a. The scale is defined by reproducible temperature points established by physical constants of readily available materials. Interpolation between these fixed points is made by platinum resistance thermometers when the temperature is below 1832°F (1000°C) and by platinum-platinum and 10% rhodium thermocouples when it is more. The National Bureau of Standards (NBS) capability for calibrating temperature-measuring devices is illustrated in Figure 1b. This figure also shows the error (uncertainty) of other thermometers at different temperatures.

ORIENTATION TABLE

The range in temperature within the universe varies from the near zero of black space to the billions of degrees in the nuclear fusion process deep within the stars. But the practical range on earth can be considered as extending from 1°R upward about five decades to around 20,000°R. This is still a tremendous range, and no single sensor could possibly cover it. Therefore, one of the restrictions on the temperature sensor concerns the temperature range over which it can stay reasonably accurate. Table A (at the front of this volume) has been prepared to show the approximate temperature ranges of each of the sensor types. The many types of sensors are listed on the left, while some of their characteristics are shown horizontally across the top. If it is not known what general type of sensor will do a specific job, the table can help point the way to the right selection.

Once the class of sensors has been found, the data in the table will give a rough idea of the applicability of that design.

When the possible choices of selection have been narrowed down to a few instrument types, the reader should turn to the corresponding sections of this volume. In the front of each section there is a summary of basic features, such as accuracy, cost, and range. Inspecting these briefly, one can determine if the instrument generally meets the requirements or not. If it does, one should read the section which describes the design and its available variations in detail. If some of the features are unacceptable, one should proceed to the next choice noted in the orientation table.

TABLE 1a
Primary Temperature Points Defined by the International Practical Temperature Scale (IPTS-90)

Equilibrium Point	*°K*	*°C*
Triple Point of Hydrogen	13.81	−259.34
Liquid/Vapor Phase of Hydrogen at 25/76 Std. Atmosphere	17.042	−256.108
Boiling Point of Hydrogen	20.28	−252.87
Boiling Point of Neon	27.102	−246.048
Triple Point of Oxygen	54.361	−218.789
Boiling Point of Oxygen	90.188	−182.962
Triple Point of Water	273.16	.01
Boiling Point of Water	373.15	100
Freezing Point of Zinc	692.73	419.58
Freezing Point of Silver	1235.08	961.93
Freezing Point of Gold	1337.58	1064.43

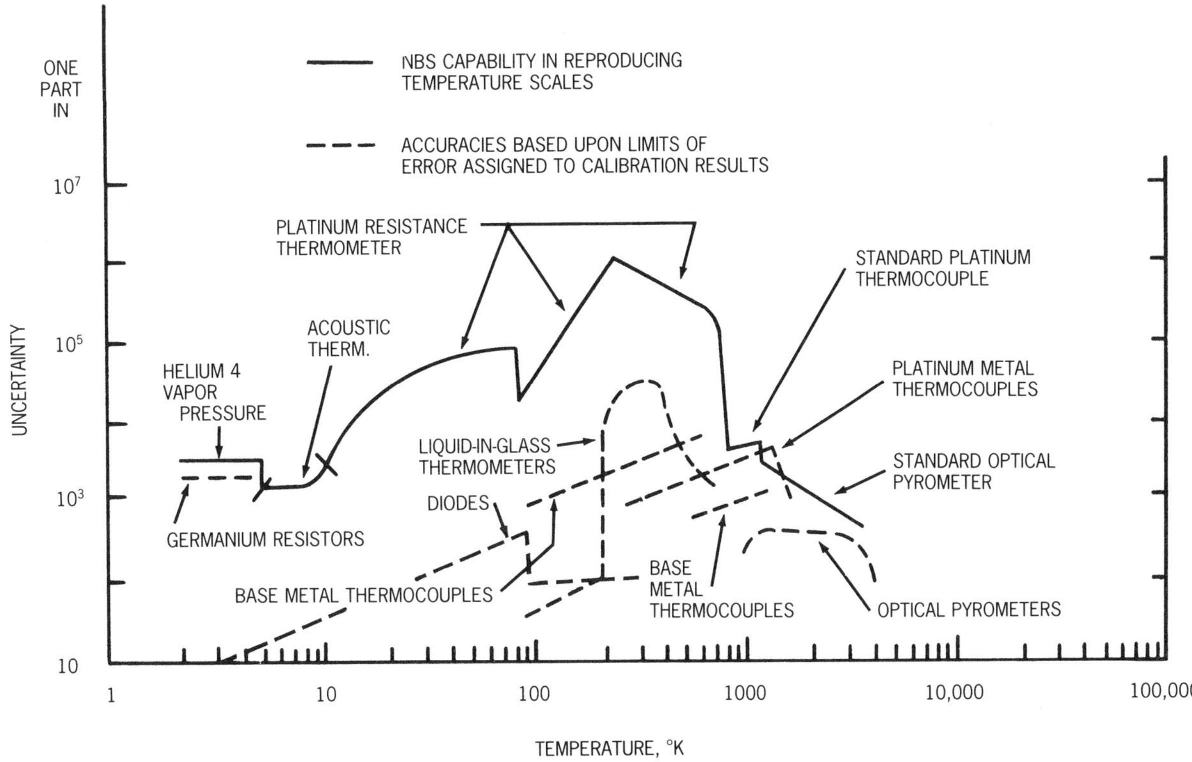

FIG. 1b
Uncertainties in calibrating different temperature sensors at various temperatures. (From NBS Technical Note No. 262)

Temperature sensors should be selected to meet the requirements of specific applications. Table 1c can assist the reader in this task.[3] If the application engineer determines the required temperature level, the nature of the information required (point or average temperature), and the nature of the process environment, this table can be used to determine the suitability of various sensors to that application. The most difficult temperature measurement applications are those where high temperatures are to be detected within a hostile environment, such as that which exists within a fluid-bed coal gasifier.

THE MANY WAYS OF MEASURING TEMPERATURE

It is believed that Galileo invented the liquid-in-glass thermometer around 1592. The principle behind the thermocouple—the existence of the thermoelectric current—was discovered in 1821 by Thomas Seebeck. The same year Sir Humphry Davy noted the temperature dependence of metals, but the first resistance temperature detector (RTD) was not constructed until 1932, by C. H. Meyers. The development of temperature sensors was a slow process until the middle of the twentieth century. Today there are nearly 20 different types of thermometers, and the old practice of using only filled thermometers or only RTDs or only thermocouples throughout a particular industrial plant is giving way to the practice of selecting each temperature sensor for a particular application, just as each level or flow detector is individually selected. This requires a better understanding of the features and capabilities of the many thermometers on the market.

The difference between the thermal expansions of two strips of metals produces the bimetallic spring, used in ranges from −80 to 800°F (−62 to 427°C) to build thermostats, bimetallic thermometers, and temperature switches. The latter can be manufactured at unit costs of a couple of dollars and serve to control various household appliances, from washing machines to dryers to dishwashers.

The other traditional thermal element, the filled system, which has a range from −300 to 1000°F (−184 to 538°C), is still used to a limited extent in HVAC (heating, ventilation, and air conditioning) and industrial applications, although there is something about nearly all of the filling materials that makes them less than ideal. Wax and other solid filling materials have just about disappeared, mercury filling is in disfavor because of its toxic characteristics, gas filling is disliked because of the requirement for a large size bulb, and vapor filling can be a problem if the process temperature crosses through the ambient. Liquid filling is still commonly used, but it also has problems—with head effects, limited capillary lengths, and compensation.

ELECTRONIC THERMOMETERS

These days the most often used industrial thermometers use sensors that change either their output voltage or their elec-

TABLE 1c
Temperature Sensor Selection Table

Measured Temperature	Under 500°C								Above 500°C							
Reading[1]	Point				Average				Point				Average			
Hostile Environment[2]	No		Yes		No		Yes		No		Yes		No		Yes	
Interference[3]	No	Yes	No	Yes	No	Yes	No	Yes	No	Yes	No	Yes	No	Yes	No	Yes
Sensors[4] Color Indicators	G(L)	G(L)	G(L)	G(L)					G(L)	G(L)	G(L)	G(L)				
Bimetallic Units	G	F	G(P)	F(P)												
Filled Elements	G	F	G(P)	F(P)	G(L)	F	F(P)	F(P)								
Resistance Bulbs	E	F	E(P)	F(P)	E	F	E(P)	F(P)	E	F	G(P)	F(P)	G(L)	F	G(P,L)	F(P,L)
Thermistors	E(L)	F	G(P)	F(P)	E(L)	F	G(P)	F(P)								
Thermocouples	G	F	G(P)	F(P)	F(L)	F(L)	F(L)	F(L)	E	F	E(P)	F(P)	F(L)	F(L)	F(L,P)	F(L,P)
Quartz Crystals	E(L)	F														
Radiation Pyrometers									E(L)	G(L)	E(L)	G(L)	E(A,L)	G(A,L)	E(A,L)	G(A,L)
Infrared Pyrometers	E(L)	E(L)	E(L)	G(L)					E(L)	E(L)	E(L)	E(L)				
Spectroscopic (Fraunhofer) Sensors															A	A
Thermopile					G(A)	F(A)							G(A)	F(A)		
Acoustic Time Domain Reflectometry (TDR)					D	D	D	D					D	D	D	D

CODE LETTERS: D-in development
L-limited
F-fair
G-good
E-excellent
P-protective well reduces speed of response
A-detects the average temperature of an area
EXAMPLE: G(D)—This combination of code letters refers to a device which is a good selection for the particular service, but is not yet commercially available.

[1]This device either detects a point or the average temperature of some section of the process or of the refractory.
[2]The term "hostile environment" here is used to mean processes such as fluid beds, where the sensor is likely to experience the mechanical impact of high velocity solid particles.
[3]"Interference" refers to need to overcome temperature interferences due to hot refractories or to temperature differences between the carrier gas and the solid particles in it.
[4]For considerations of measurement error, span, cost stability, response time, linearity, materials of construction, etc., refer to the text.

tric resistance as a function of temperature (Figure 1d). Of these, probably the most popular are the thermocouples (TC). They have a wide temperature range (−440 to 5000°F, or −262 to 2760°C), relatively low cost, and many physical forms, making them highly versatile. When used on high-temperature combustion processes, it is necessary to continuously detect the loop resistance of the thermocouple, because a sudden drop in that resistance signals thermocouple burnout.

The main limitation of thermocouples is the small amount of voltage that is generated by a unit change in temperature. For example, a one degree Fahrenheit temperature change on a platinum thermocouple results in an output voltage change of 5 microvolts. In contrast, some of the temperature transmitters might have a minimum span of 3 millivolt and an inaccuracy of 0.25% of that full range, or an error of 7.5 microvolts. Consequently, thermocouples are not recommended for use on narrow span or small differential temperature measurement applications.

Similarly, because of the weak output signal, multiplexer contact resistances (over 1 microvolt) can also contribute relatively large errors. Because the thermocouple generates the smallest signal of all thermometers, it is the most affected by noise and lead-wire errors. Because the thermocouple is a temperature difference sensor, it is also subject to cold junction compensation and lead-wire junction errors. Most of these problems can be overcome by placing the converter electronics directly on top of the thermowell and thereby eliminating the problems associated with the transmission of the weak thermocouple signal.

Resistance temperature detectors (RTDs) have a narrower range, from −430 to 1200°F (−257 to 649°C), but otherwise are equally as popular as the thermocouples. They are more linear and more stable and can accurately measure smaller temperature spans or differentials than can the thermocouples. As with thermocouples, it is recommended to mount the electronics directly on top of the RTD and thereby eliminate the need for lead wires. If this is not done, the resistance of equal-length lead wires can still differ by 10%, and even the three-wire installations can cause substantial errors. (A standard 100 ohm platinum RTD changes its resistance by 0.385 ohms/°C, while if it has 500 ft long leadwires, they can be mismatched by a full ohm.) Other potential error sources include the aging of reference resistances, change in the voltage generated by the power supply, and moisture penetration.

Thermistors are miniature metallic oxide elements, which generate the most sensitive and the fastest response of any of the thermometers. For each degree Centigrade, their resistance changes by about 200 ohms. Their range is only from −150 to 600°F (−101 to 316°C) and their output is very nonlinear. Because the thermistor element usually has a

FIG. 1d
The characteristics of electronic thermometers.

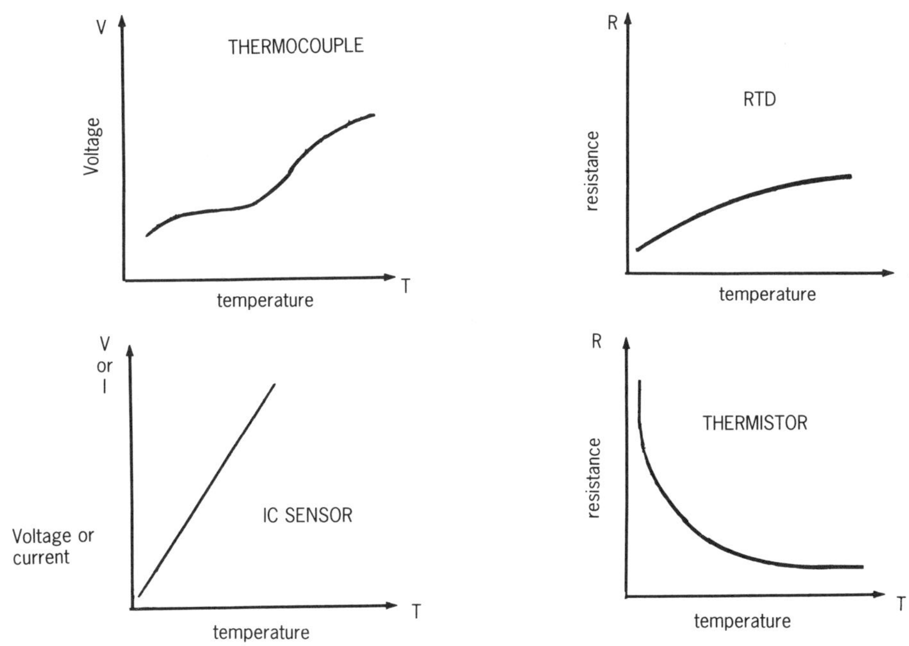

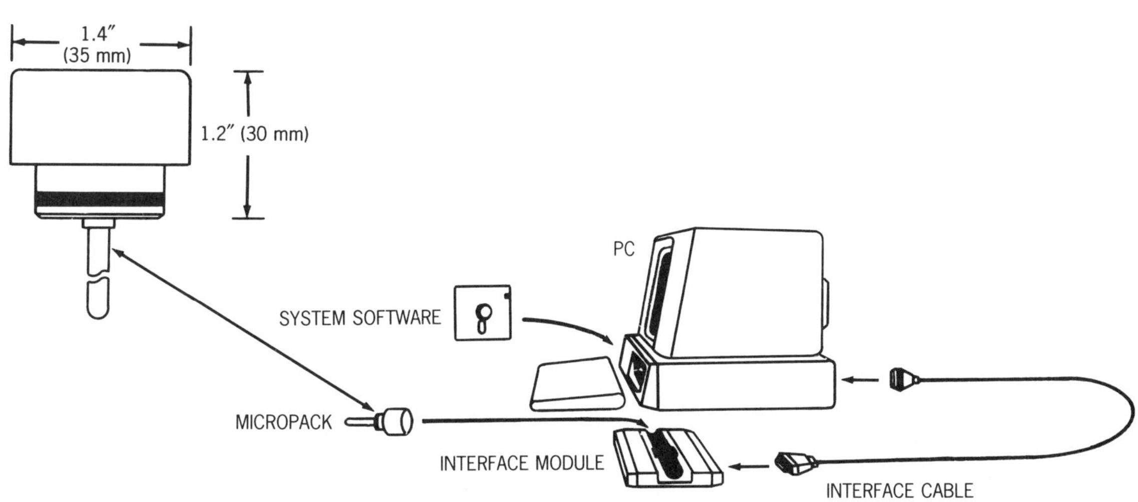

FIG. 1e
Thermistor-based micropac element can travel with product and provide time/temperature profiles. (Courtesy of Ball Corporation, Industrial Systems Div.)

TABLE 1f
Temperature Sensor Accuracy and Range

	Description of Sensors	*Temperature Range**	*Inaccuracy (Error)* — *ISA Standard Limits*	*Inaccuracy (Error)* — *ISA Special Limits*
I	Thermocouples			
	a. Chromel/Alumel (K)	−300°F to +2500°F	±4°F, 0°F to +530°F	—
			±¾%, +530°F to +2500°F	
	b. Iron/Constantan (J)	−310°F to +1400°F	±4°F, 0°F to 530°F	±2°F, 0°F to +530°F
			±¾%, +530°F to +1400°F	±⅜%, +530°F to +1400°F
	c. Copper/Constantan (T)	−310°F to +750°F	±2%, −150°F to −75°F	±1%, −300°F to −75°F
			±1½°F, −75°F to +200°F	±¾°F, −75°F to +200°F
			±¾%, +200°F to +700°F	±⅜%, +200°F to +700°F
	d. Platinum (S) 10% Rhodium/Platinum	+32°F to +3200°F	±5°F, +32°F to +1000°F	—
			±½%, +1000°F to +2700°F	
	e. Tungsten/Tungsten 26% Rhenium	+32°F to +5200°F	±8°F, +32°F to +800°F	—
	f. Tungsten 5% Rhenium/Tungsten 26% Rhenium	+32°F to +5200°F	±1%, +800°F to +4200°F	—
II	Resistance Thermometers			
	a. Platinum (high-purity strain-free element)	−435°F to +900°F	±0.01°F limited range, ±0.05% full range	
	b. Platinum (commercial purity element)	−320°F to +2000°F	±0.2% full range	
	c. Nickel	−320°F to +800°F	±0.4% full range	
	d. Balco	−320°F to +500°F	±0.4% full range	
	e. Semiconductor (germanium)	−458°F to −400°F	±0.01°F	
	f. Semiconductor (thermistors)	−40°F to +300°F	±0.05°F limited range, ±1°F full range	
III	Bimetallic Thermometers	0°F to +1000°F	±2°F limited range, ±10° full range	
IV	Liquid-Filled Thermometer (glass stem)	−328°F to +800°F	±0.05°F limited range, ±1° full range	
V	Liquid-Filled Thermometer (metal bulb)	−150°F to +700°F	±2°F limited range, ±5°F full range	

*To convert from °F to °C use the following equations: $+°C = \frac{5(°F - 32)}{9}$ and $-°C = \frac{5(°F + 32)}{9}$

resistance of about 5000 ohms, even small current flows can cause self-heating.

Thermistors can be packaged in a small, portable element, which can be inserted into the process and can memorize the process temperature while it is traveling with the product. When this temperature memory package emerges at the discharge point, it can be plugged into and interrogated by a computer to thereby obtain the full temperature history of the process (Figure 1e).

Integrated circuitry sensors have a limited temperature range (67 to 257°F, or −55 to 150°C) and are not yet available in the variety of packaging that thermocouples and RTDs are. Their main advantages are their low costs (some under $10) and their linear and strong output signals. Diodes are mostly used in the cryogenic temperature range of −271 to 202°C (−455 to 396°F) and when calibrated can be accurate to 0.05 to 0.1°C.

NONCONTACT PYROMETERS

In combustion processes, the temperature can be measured on the basis of the radiation emitted by the hot body. Total radiation pyrometers operate by measuring the total amount of energy radiated by a hot body. Their temperature range is from 0 to 7000°F (0 to 3890°C). These pyrometers have been gradually replaced by infrared (IR) pyrometers, which detect the dominant wavelength of the radiation received from a hot object. The basis of infrared thermometry is the fact that as temperature increases, the dominant wavelength of hot body radiation gets shorter. Infrared thermometers have about the same range as total radiation ones.

Newer developments in temperature pyrometry involve the automatic measurement of changing surface emissivities of hot objects, using laser beams to automatically correct these changes. The other major development is the coupling of infrared pyrometry with fiber-optic (FO) technology, which can focus the temperature measurement on inaccessible points in unfriendly environments that are a long distance from the optical/electronic readout instruments. Some of these fiber-optic thermometers are capable of high accuracy, about 0.1% at around 1000°C, and can operate over a range of from 932 to 3632°F (500 to 2000°C). Fiber-optic systems can also be multiplexed so that the cost of expensive electronics is shared between several temperature measurements.

Noncontacting temperature measurement can also be achieved by the measurement of radiated heat using thermopiles (Figure 1j). These devices are unaffected by color, finish, or emissivity changes as they depend only on the convection of heat.

SELECTION OF RANGE AND ACCURACY

The Orientation Table at the front of this volume (Table A) and Table 1f both provide information on the ranges and accuracies of the various types of thermometers. For example, if the temperature to be measured is around 2000°F (1143°C) that requirement alone narrows the choice to either pyrometers or thermocouples. On the other hand, if the temperature to be measured is 200°F (93°C), just about all the temperature sensors can be considered and therefore one can take other factors, such as size, speed of response, linearity, or cost into consideration. The factor most often considered right after range is accuracy.

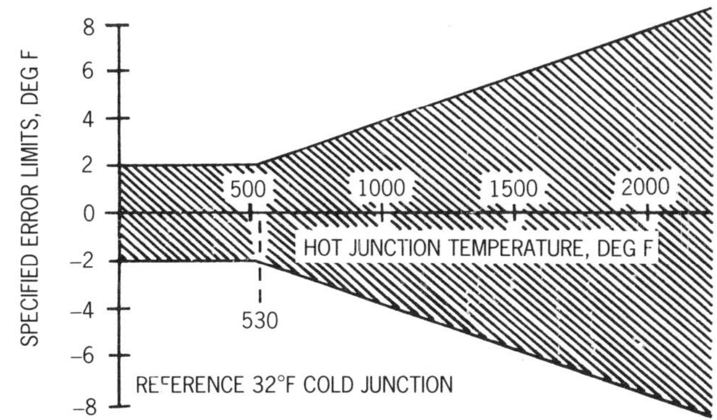

FIG. 1g
The error limits of an iron-constantan thermocouple manufactured to meet the "special" limits of ISA.

Figure 1g illustrates the error limits of an iron-constantan thermocouple wire element if it is manufactured to meet the "special" quality requirements of the Instrument Society of America (ISA). In making a selection, one should evaluate the different sensors considered against the actual accuracy requirement of the installation. It should also be recognized that the total error of the temperature measurement or control system will be more than just the error contribution of the sensing element. Added to this error will be the inaccuracies contributed by the installation (lead wires, junctions, noise, etc.) and by the converter, transmitter, and readout devices.

For example, the typical error contribution of an integral thermocouple transmitter is 0.1% of span or 5 microvolts, whichever is larger (in the case of platinum thermocouples, this amounts to an error of about one degree Fahrenheit). For integral platinum RTD-type transmitters the error is usually 0.15% of span or 0.15°F (0.08°C), whichever is larger, and for nickel RTDs it is 0.25% of span. The best obtainable digital-to-analog (D/A) conversion error is 0.025% of the span. In addition, supply voltage variation can contribute 0.02% of span and ambient temperature changes can cause additional zero, span and reference junction errors.

HIGH-TEMPERATURE SERVICE

At high temperatures in the 2000 to 3000°F (1100 to 1650°C) range, the problem of thermowell drooping becomes a problem when using thermocouples. If direct temperature measurement is to be used (where the temperature sensor is physically inserted into the process), metallic thermowells are likely to droop, resulting in problems of replacement and maintenance. If ceramic thermowells are used instead of metallic thermowells, erosion becomes a problem in fluid bed-type environments due to the relatively low surface strength of the ceramic materials.

If the thermowell is flush with the inside diameter of the reactor, its life is likely to be increased, but only at the expense of having a less representative temperature measurement because of the greater influence of wall temperatures. Another way of increasing the reliability of the overall installation is to provide automatic, scheduled, preventive maintenance replacement of the thermowells. This can be done on a redundant basis, so that at any one time two or even three temperature sensors are in operation. A voting system can be arranged in critical locations so that the reading of the one sensor which disagrees with the other two is disregarded. While it is desirable to replace the thermowells without shutting the process down, and while it is desirable to replace them before they fail, all this effort is not inexpensive, and over a period of years of operation it can easily approach the initial cost of noncontact sensors, such as infrared pyrometers.

The trend in high-temperature measurement favors the indirect, noncontacting-type measurements. The unit cost of such installations can be lowered by making multiple readings, using fiber optics to connect many sensing points to the same infrared source and the same receiving electronics package. This seems to be the most promising solution to the problems of high-temperature measurements in hostile environments.

MEASURING THE TEMPERATURE OF SOLIDS

Determination of the allowable size and configuration of the sensor requires some knowledge of the heating or cooling conditions together with an estimate of the magnitude of the temperature gradients that are likely to exist in the region in which the measurement is to be made. A simple rule-of-thumb indicator to determine if significant gradients are likely to be present is the magnitude of the Biot modulus (hL/K), where h is the surface heat transfer coefficient, L is the smallest dimension of the solid, and K is the thermal conductivity of the solid. If this modulus is over 0.2, significant temperature gradients are likely to exist in the solid, and care should be exercised in choosing the size, location, and orientation of the sensor within the solid. If the Biot modulus is less than 0.2, no significant gradient is expected and a measurement anywhere on or within the solid should give identical results regardless of size or configuration of the sensor. If significant gradients are likely to exist, the maximum rate of heat transfer to the surface of the solid must be known or estimated, and the maximum gradient at the point of measurement must be determined. The following relationship allows the maximum gradient at the surface of a solid to be calculated:

$$\frac{\Delta T}{\Delta X} = \frac{q}{K} \qquad \mathbf{1(1)}$$

where

$\frac{\Delta T}{\Delta X}$ = temperature gradient at the surface
q = heat transfer rate per unit area at the surface
K = thermal conductivity of solid

Under certain conditions of heating or cooling, if measurements at points other than the surface are important, it may be necessary to evaluate anticipated heat transfer conditions and resulting temperature gradients.[4,5]

On the basis of this gradient, it is possible to establish limits on the size of the sensing device. For example, the length of any one of the three dimensions of the sensor (lead wires excluded) should not be greater than the distance between two points of the process that are different in temperature by more than the acceptable measurement error. It is assumed that the sensors are in satisfactory thermal coupling with the process material, which is not always the case. If the thermal coupling is poor, the sensor will not reflect the true temperature history experienced by the solid, a condition which can produce dynamic errors.[6] The best thermal coupling is achieved by direct bonding of the sensor, such as welding a thermocouple to the solid surface or into a cavity within the solid. The bond line between the sensor and the solid should be kept as thin as possible and should not fracture or fail during thermal cycling. Various epoxy and ceramic cements, with fillers to improve their conductivity, have been successfully used for such bonding. For example, a flat resistance thermometer bonded to a surface with an epoxy bond line 0.005 in. (0.1 mm) thick will produce a lag time of about 1 second, which will produce a dynamic error equal to the rate of temperature rise of the surface, i.e., a dynamic error of 25°F (14°C) for a rate of surface temperature change of 25°F per second.[6]

Methods of installing temperature sensors on or within solid bodies are shown in Figure 1h.

MEASURING THE TEMPERATURE OF FLUIDS

The fundamental problem of measuring the temperature of a fluid is one of assuring strong thermal coupling. For a fluid temperature measurement to have meaning, the sensor must come to equilibrium with the temperature of the fluid. The difference between the equilibrium temperature of the sensor and the fluid temperature is a direct error. Consequently, with rapidly changing temperatures the rate of heat transfer between the sensor and process fluid must be sufficient to overcome the thermal capacity of the sensor in order that it can follow the fluctuations in fluid temperature. Under such conditions one can write the following equation which ex-

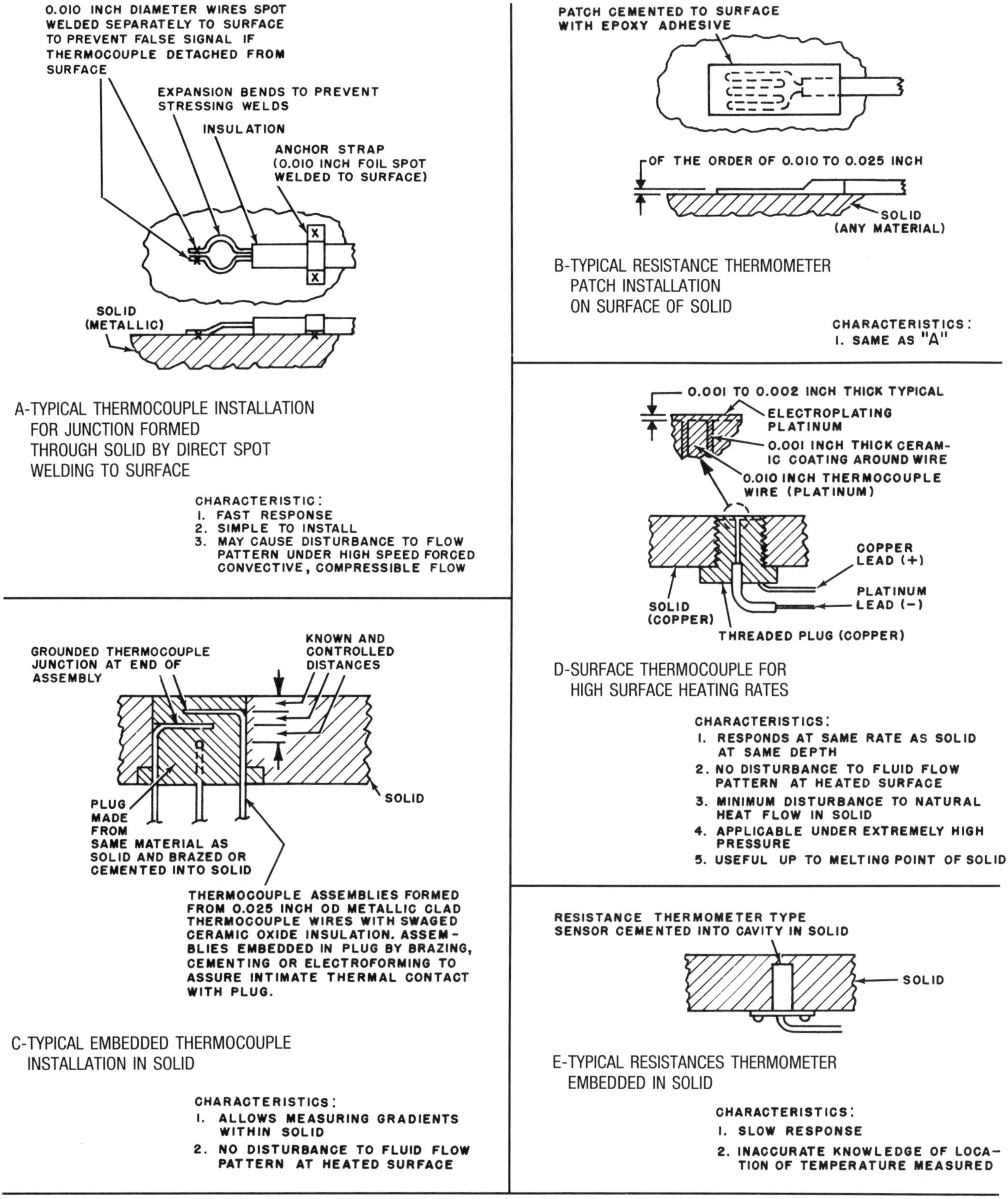

FIG. 1h

Some selected methods of installing temperature sensors on or within solids.

presses the temperature, T, experienced by a sensor which was initially at temperature T_1:[4]

$$\frac{T_1 - T}{T_1 - T_0} = e^{-(hA/WC_p)\theta} \qquad 1(2)$$

where

T = temperature
e = base of natural logarithm
h = heat transfer coefficient at surface of sensor
A = surface area of sensor
W = weight of sensor
C_p = specific heat of sensor
θ = time after immersing sensor in fluid

A sensor, to respond rapidly to changes in fluid temperature, should have a large surface-area-to-mass ratio. Furthermore, the heat transfer coefficient, which is a direct function of the fluid mass flow rate over the sensor, should also be as large as possible. Forced convection or rapid flow of the fluid over the sensor is therefore desirable.

A practical measure of sensor sensitivity is its time constant. Since the sensor approaches change in fluid temperature asymptotically, it is difficult to determine when it has reached that temperature. The time constant of a sensor is the time required to accomplish 63.2% of the step change. This point is reached when the exponent of Equation 1(2) is 1. Thus, the time constant of a fluid stream temperature sensor (τ) is defined as:

$$\tau = \frac{1}{hA/WC_p} \qquad 1(3)$$

When measuring the temperature of a gas stream in a heated duct or furnace where temperature differences between the sensor and its surroundings exceed 1000°F (538°C), significant errors can occur due to radiation exchange between the sensor and its surroundings. Under such conditions the sensor must be shielded against thermal radiation exchange. This can cause disturbances in the flow of the fluid around the sensor and hence affect the directional response characteristics.

Another thermal effect in measuring the temperature of gas streams at high velocities is the recovery factor, which results from an increase in the temperature of the gas at the sensor due to compression heating as the gas is brought to stagnation against the sensor.[7,8] Fluid stream sensor configurations are shown in Figure 1i together with some installation and application guidance.

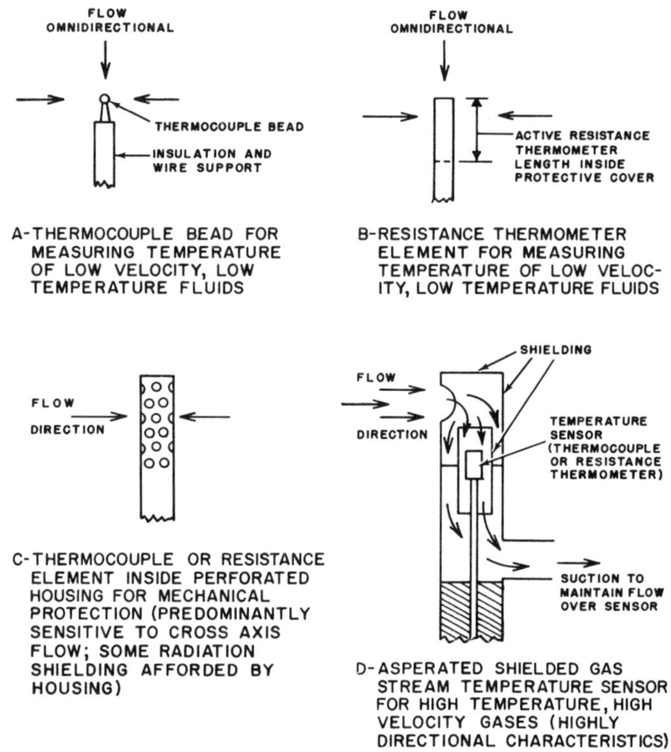

FIG. 1i
Fluid stream sensor configurations.

ATMOSPHERIC EFFECTS

Adverse atmospheric conditions can cause problems with the measurement of temperature. For example, in a highly humid or very moist environment it is essential that the element of a resistance thermometer or the bead of a thermistor be well insulated electrically. If moisture contacts the resistance element or thermistor bead, it may cause a short. Thermocouples in general are less sensitive to moisture than are resistance thermometers. If null balance potentiometric or high input impedance readout devices are used, insulation resistance between legs of the couple as low as 10,000 ohms can be tolerated without serious error in the indicated temperature. If low input impedance current measuring readout devices are used, a high insulation resistance between legs of the couple becomes as important as with resistance thermometers and should be in the megohm range.

Corrosive, reducing, or oxidizing environments also create problems. If a sensing device is exposed to such environments, it must be protected by some form of envelope or coating. It should be remembered that protection applied to a sensor will increase its mass and thus, in general, adversely affect its response characteristics.

TEMPERATURE OF MOVING SURFACES

Measuring the surface temperatures of webs of paper, plastics, textiles, or metals or of rotating cylinders (such as calendar rolls, rotary kilns, or drier cans) requires special thermometer designs. These instruments are either noncontacting or traveling-brush-skid designs (see Figure 1j). Some use thermocouple elements, while others use RTDs or even bimetallic springs. Still others are of the conventional or fiber-optic infrared variety or of the convective null-heat-balance design. Figure 1j illustrates some of the thermocouple-based designs which, while somewhat fragile and sub-

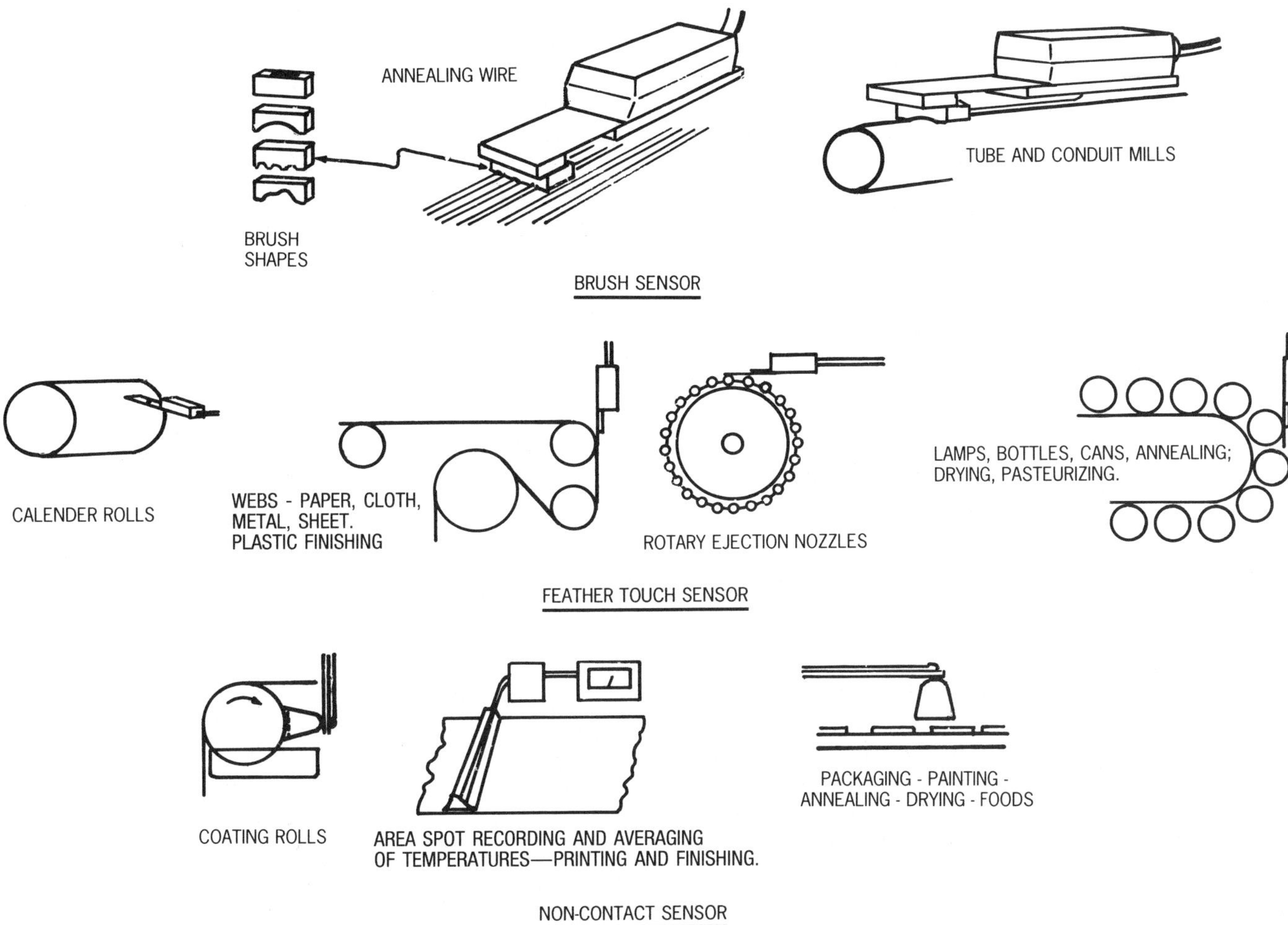

FIG. 1j
Thermocouple-type surface temperature sensor designs. (Courtesy of Syscon International Inc.)

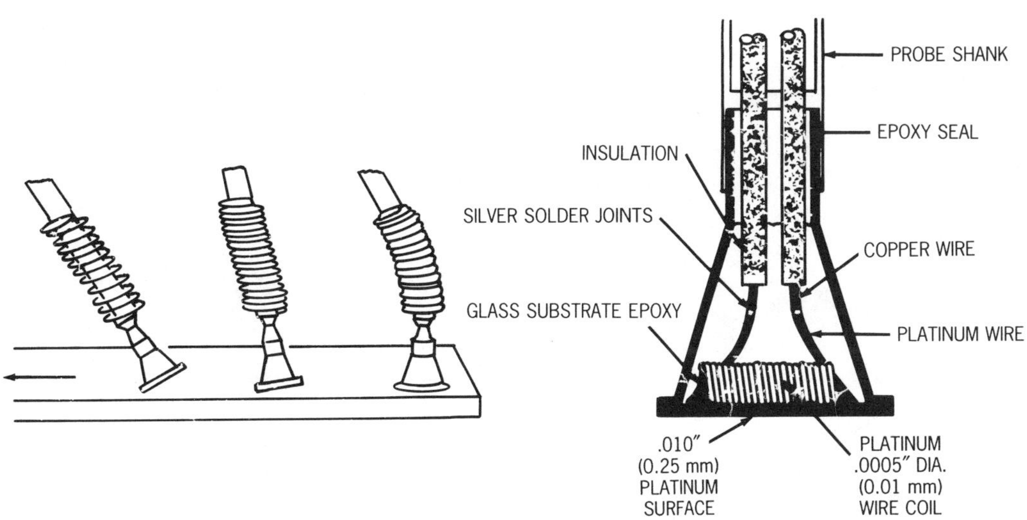

FIG. 1k
RTD-based surface-temperature detector. (Courtesy of Wahl Instruments Inc.)

ject to ambient errors, have still gained reasonable acceptance. The brush design is available for use from −300 to 1000°F (−184 to 538°C) and the contact bar can be lead, copper, brass, or mild steel. The feather-touch units operate up to 1600°F (870°C) in intermittent and up to 1000°F (538°C) in continuous services. The noncontacting designs use type J thermocouples, are constructed mostly of stainless steel, and can be used from −150 to 1000°F (−100 to 538°C).

Figure 1k illustrates a surface-temperature detection probe consisting of a platinum RTD and stainless steel housing. The coil spring design serves to bring the 0.25 in. (6 mm) diameter sensor tip in uniform contact with the surface being measured.

Depending on the environmental conditions, radiation pyrometers can also be used to detect the temperature of moving surfaces (Figure 1l). The portable units are convenient for maintenance purposes, such as the detection of evolving shorts in cables, the spotting of leaking steam traps or insufficient thermal insulation, and so on (see Figure 1m). The main limitations of IR thermometers are caused by changes in the emissivity of the surface and by contamination and dirt buildup. Laser-based compensation is claimed to overcome the emissivity problems, and purging can be useful to reduce contamination in fiber-optic infrared thermometers (Figure 1n). The radiation pyrometers are usually either portable or permanently installed infrared thermometers. When the temperature of a hard-to-reach area is required (say the detection of the surface temperature of turbine blades inside combustion turbine engines), the fiber-optic infrared sensors are likely to be used.

The convective null-heat-balance concept[9,10] involves the null-balance of convective heat transfer between a sensing head of known temperature and a surface of unknown temperature, in which temperatures are compared rather than directly measured. If, for example, two surfaces are placed in close proximity to one another, their temperatures can be compared by measuring the rate and direction of heat transfer between them. The direction of heat transfer is a definite measure of which surface is at the higher temperature; if the heat transfer between them is zero, their temperatures must be equal.

If one of these surfaces contains a heat transfer sensor and its temperature is known (T), one can readily determine if the temperature of the other surface is either below, equal to, or above T by simply observing whether the heat transfer is zero or not and what its direction is. It is not necessary to measure the magnitude of the heat transfer. The technique, of course, works only if the point at which heat transfer is being sensed is isolated from surrounding influences. On a flat surface, isolation is accomplished by locating this point at the center of a relatively large isothermal plate (typically

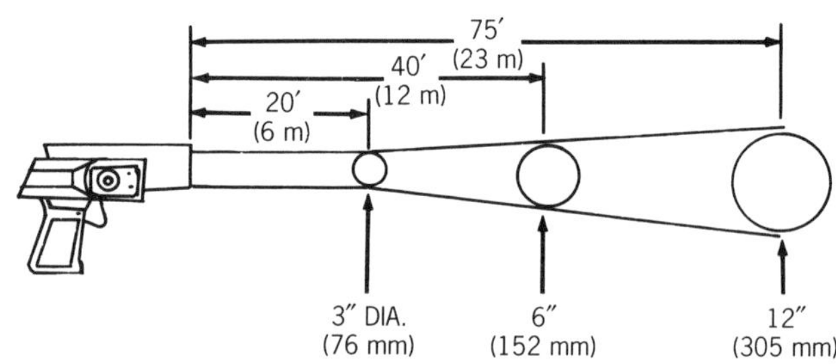

FIG. 1m
Typical target sizes and distances of portable infrared thermometers. (Courtesy of Wahl Instruments Inc.)

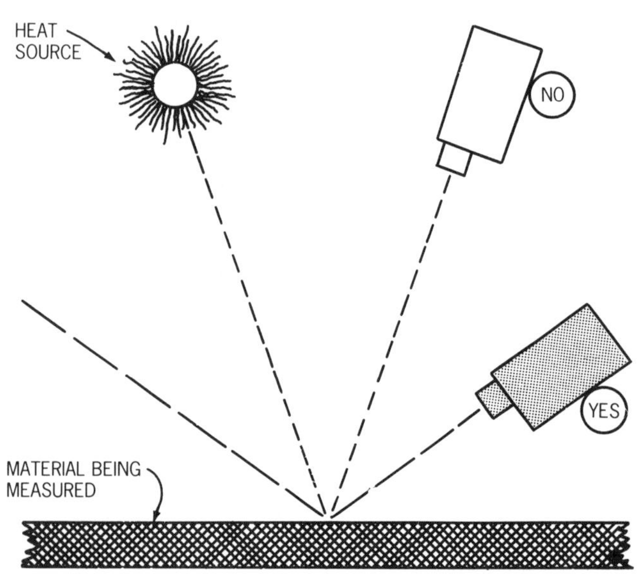

FIG. 1l
The infrared thermometer must be carefully aimed to avoid focusing on reflected solar radiation.

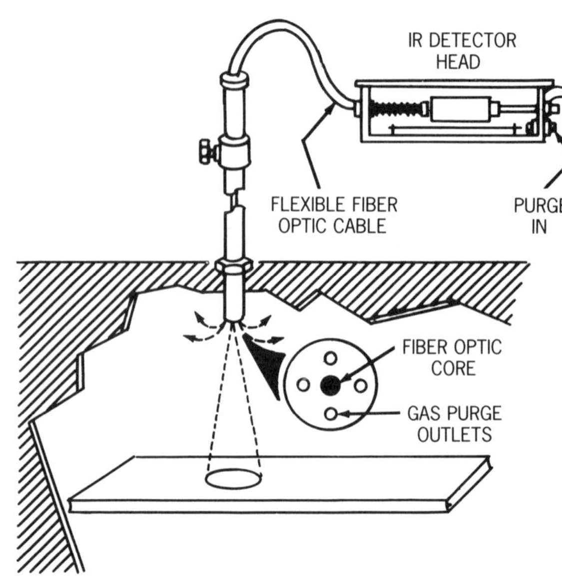

FIG. 1n
Purged fiber-optic elements can detect the temperature of small, moving surface areas in hostile environments such as furnaces. (Courtesy of Vanzetti Systems Inc.)

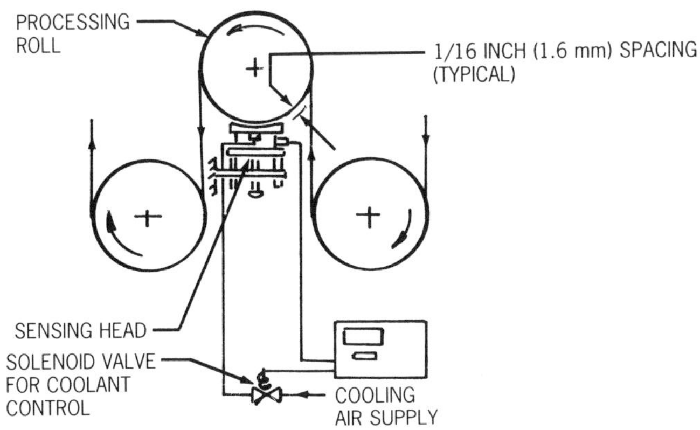

FIG. 1o
Noncontact thermometer based on convective null-heat-balance. (Courtesy of Trans Met Engineering Inc.)

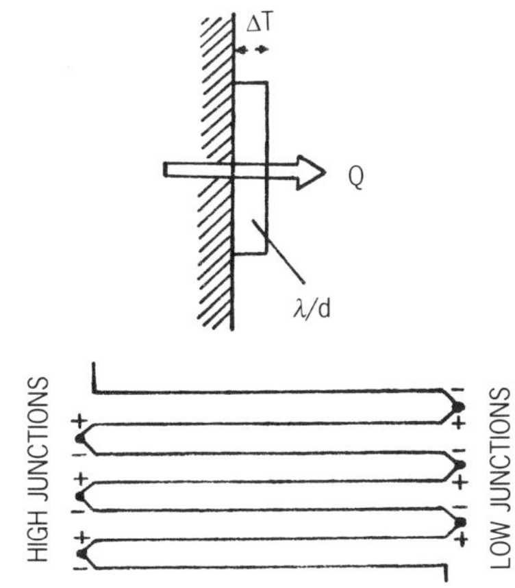

FIG. 1p
Knowing the thermal conductivity (λ) and the sensor plate thickness (d), the measurement of temperature difference (ΔT) can be converted into heat flow (Q), because Q = (λ/d) ΔT.

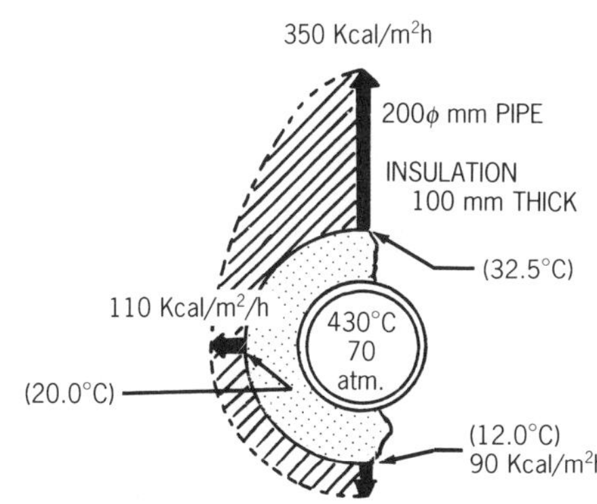

FIG. 1q
The measurement of surface heat loss from an insulated steam pipe. (Courtesy of Showa Denko K.K.)

6 in./150 mm in diameter) spaced approximately 1/16 to 1/8 in. (1.5 to 3 mm) from the surface being measured.

The basic concept of null-heat-balance can be used either to monitor or to control surface temperature. In the monitoring mode, the temperature of a sensing head is automatically varied to keep it at null-heat-balance (at the same temperature) as the moving surface. The head temperature, which is equal to the surface temperature, can then be measured (Figure 1o).

In the control mode, the head is held at the desired setpoint temperature at which the surface is to be controlled and the signal from the heat flow sensor is employed to manipulate the surface temperature.

In another arrangement, the signal from the heat flow detector is calibrated for temperature difference between the sensor, which is at a constant known temperature, and the surface being measured.

The convective null-heat-balance-type surface thermometers can detect temperatures up to 500°F (260°C) and can follow changing temperatures up to the rate of 15°F/min (8.3°C/min); their measurement error is about 2% of full scale.

HEAT-FLOW AND THERMAL-CONDUCTIVITY SENSORS

The accurate measurement of heat flow through thermal insulators and of the thermal conductivity of construction materials are both important. Such measurements are of interest not only to architects, but also to engineers who are involved with safety and energy conservation.

A common heat flowmeter design involves the placing of a thin plate of known thermal conductivity on a heat radiating surface. It has been found that the heat flow through these elements is directly related to the temperature difference through them. This temperature difference is often detected by thermopiles, a large and even number of thermocouples connected in series in such a manner that their high-temperature junctions are on the inside and their low-temperature junctions are on the outside surface of the sensing element (Figure 1p). The heat flows that are encountered in different processes range from about 10 k-cal/m^2 hr through freezer walls to about 100,000 k-cal/m^2 hr through the shells of water-cooled electric furnaces. The thickness of the sensor plates is a few millimeters, and the plates are made out of rubber, organic material, or other heat-resistant materials, sometimes contained in a thin, stainless steel disk case.

The heat flow distribution frequently varies with the direction of heat flow. Figure 1q, for example, shows that the heat loss from the top of a steam pipe was found to be much more than through the bottom surface. Such findings are usually explained by noting that heat flow from a surface is not only a function of the surface temperature, but also of the effects of coating. It has been found that if a surface is coated with bright and glistening aluminum paint, it will radiate much less heat at the same temperature as a surface where the coating has worn off. Sensor elements can measure the variations in heat flow at different points on many shells, from LPG tank walls to electric and blast furnace shells, and

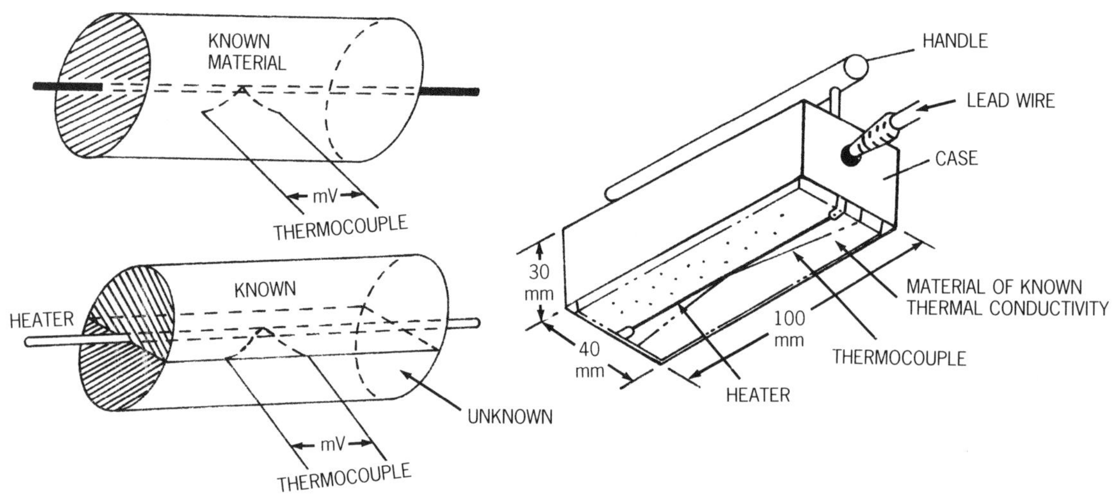

FIG. 1r
The design of a thermal conductivity sensor. (Courtesy of Showa Denko K.K.)

their readings can reveal the erosion of linings, as well as other hard-to-detect phenomena.

In other processes the interest is in measuring the thermal conductivity of heat insulating substances, preferably with a quick (one-touch), nondestructive method. The thermal conductivity of these substances ranges from 0.01 to 10 k-cal/mhrC. The hot wire method of thermal conductivity measurement involves the stretching of a thin heating wire through a sample and applying a constant amount of power (Figure 1r). The higher the thermal conductivity of the sample, the lower will be the resulting surface temperature of the heater wire. Therefore, it is possible to read the surface temperature of the heater wire and interpret from that reading the thermal conductivity of the sample.

If the surface temperature response curve (time versus temperature) of the wire is recorded when it is surrounded by a known thermal conductivity material and then half of the known sample is replaced by a material having an unknown conductivity, the difference in the response curves, when the test is repeated, can be correlated to the thermal conductivity of the unknown substance. In the sensor shown in Figure 1r, the heated wire is surrounded on the top by a material of known thermal conductivity; therefore, when it is placed on the flat surface of the unknown sample, its temperature response curve will reflect the thermal conductivity of that substance.

INTELLIGENT TEMPERATURE TRANSMITTERS (ITT)

The incorporation of a microprocessor has improved the performance of temperature transmitters, just as it has improved the performance of all other devices. These improvements include the near elimination of errors due to ambient temperature variations. In the case of RTD elements, a 50°F (28°C) change in the ambient results in 0.1% error in the input ohms or in a 0.2 ohm error, whichever is greater. In case of TCs, the same ambient change results in 0.1% error in the input millivolt value or in a 5 microvolt error, whichever is greater. The same ambient change results in a cold junction error, which is about 1/50th of the ambient change. The effect of supply voltage variation is reduced to about 0.005% of span per volt.

Some ITTs store their recalibration reference values within the microprocessor and perform 3-point self-calibration every 5 seconds. Other ITT units provide melt/freeze references for single-point on-line calibration. Isothermal Technology Ltd. of England inserts near the sensor a capsule of pure metal that has a melting point below the operating temperature of the process. During shutdown or start-up a calibration check is performed, as the melting or freezing of the capsule temporarily arrests the change in temperature.

An ITT usually provides a linear output with all its sensing elements and can also be reconfigured to accept any thermocouple, RTD, or millivolt sensors. The reconfiguration process is quick and convenient, and it tends to allow for lower inventories by making the transmitters interchangeable. The ITT will also automatically select the right cold junction compensation to match the selected thermocouple. Some of the ITT designs are capable of handling dual RTD elements. This allows for temperature averaging, temperature difference measurement, or automatic RTD sensor switchover if the primary sensor fails in a redundant installation.

Communications with modern ITT units does not interrupt their measurements, and their remote diagnostics capability simplifies troubleshooting and increases their availability. The ITT units are available in explosion-proof housings, and some of them can also be provided with displays in units of Fahrenheit, Celsius, Rankine, or Kelvin (°F, °C, °R, or °K). Changing zero or span is just as easy as changing the type of sensor or the engineering units of display.

It is expected that ITTs will also contribute in the future to improved sensor maintenance. In the case of thermocouples, diagnostic capability can be provided to detect the degradation of the thermal insulation by either measuring the wire-to-sheet capacitance or by measuring its resistance; similarly, junction deterioration can be detected by measuring loop resistance. Temperature sensor time constants can be detected by applying self-heating.[11]

In the case of platinum RTDs the diagnostics might also include the measurement of insulation resistance. Because most RTDs are provided with three or four wires (two leads at one or both ends of the element), the presence of thermocouple junction effects can be checked and measured by reversing the leads if DC sensing current is used. Similarly to thermocouples, Joule heating can also be used with RTDs to determine sensor response time.[11]

Further advances can occur if some of the thermocouple/RTD sensors are modified or redesigned. For example, it would seem feasible to incorporate both an RTD and a TC element within the same sheet and periodically crosscheck their readings. This would improve their reliability as it is unlikely that both would degrade at the same time. Multisensors could also be used to improve sensor response by arriving at the compensated output signal as twice the reading near the tip minus the reading further away from it. It seems that diagnostics could also be improved if instead of units with two thermocouple wires, multi-lead units were used.[11]

References

1. Stimson, H.F., ''International Practical Temperature Scale of 1948, Text Revision of 1960,'' NBS Monograph 37, 1961.
2. NBS Special Publication 300, Volume II, January 1969.
3. Lipták, B.G., ''Overview of Coal Conversion Process Instrumentation,'' Argonne National Laboratory publication: ANL-FE-49628-TM01, May 1980.
4. Giedt, W.H., *Principles of Engineering Heat Transfer,* New York: Van Nostrand, 1957.
5. Carslow, H.S., and Jaeger, J.C., *Conduction of Heat in Solids,* London: Oxford, 1947.
6. Rall, D.L., and Hornbaker, D.R., ''A Rational Approach to the Definition of a Meaningful Response Time for Surface Temperature Transducers,'' New York: 21st Annual ISA Conference, October 22–27, 1966.
7. Glawe, G.E., Simmons, F.S., and Stickney, T.M., ''Radiation and Recovery Corrections and Time Constants of Several Chromel-Alumel Thermocouple Probes in High-Temperature, High-Velocity Gas Streams,'' NACA TN-3766, October 1956.
8. Scadron, M.D., and Warshawsky, I., ''Experimental Determinations of Time Constants and Nusselt Numbers for Bare-Wire Thermocouples in High-Velocity Air Streams and Analytic Approximations of Conduction and Radiation Errors,'' NACA TN-2599k, January 1952.
9. U.S. and foreign patents have been issued and others are pending, Trans-Met Engineering, Inc., La Habra, California.
10. Hornbaker, D.R., and Rall, D., ''The Convective Null-Heat-Balance Concept for Non-Contact Temperature Measurements of Sheets, Rolls, Fibers and Wire,'' Fifth Symposium on Temperature—It's Measurement and Control in Science and Industry, June 21–29, 1971, Washington, D.C.
11. Kerin, T.W., and Katz, E.M., ''Temperature Measurement in the 1990's,'' *InTech,* August 1990.

Bibliography

Bediones, D., and Wang, T.P., ''Criteria for the Selection of Thermocouples Versus RTD's in Industrial Applications,'' 1991 ISA Conference, Paper #91-0300.

Bidini, G., ''A Method for the Thermo-Fluid Dynamic Design of Gas Turbine Total Temperature Probes,'' 1991 ISA Conference, Paper #91-0511.

Bliss, P., and Morgan, R.K., ''Errors in Temperature Measurement,'' *Instrumentation Technology,* March 1971.

Bonkowski, ''Infrared Thermometry,'' *Measurements and Control,* February 1984.

Carlson, D.R., ''Temperature Measurement in Process Control,'' *InTech,* October 1990.

Carlson, D.R., ''Transmitting a Remote Temperature,'' *Control,* April 1989.

Hall, J., ''Applying Temperature Sensors,'' *Instruments and Control Systems,* June 1980.

Hashemian, H.M., and Peterson, K.M., ''Assuring Accurate Temperature Measurement,'' *InTech,* October 1989.

Hormuth, G.A., ''Ways to Measure Temperature,'' *Control Engineering,* Reprint No. 948, 1971.

Jutila, J.M., ''Temperature Instrumentation,'' *Instrumentation Technology,* February 1980.

Kreider, K., ''Thermocouple Research at NIST,'' 1991 ISA Conference, Paper #91-0301.

Krigman, A., ''Temperature Instrumentation: Recent Patent Activity,'' *InTech,* August 1983.

Langan, P.E., and Ironside, D.S., ''Calibration Ensures Accurate Temperature Measurement,'' *InTech,* October 1989.

Leewis, W., ''The International Temperature Scale of 1990,'' 1991 ISA Conference, Paper #91-0302.

Lipták, B.G., ''Overview of Coal Conversion Process Instrumentation,'' Argonne National Laboratories Publication: ANL-FE-49628-TM01.

Peacock, G.R., ''Radiation Thermometry: Temperature Measurement Without Contact,'' 1991 ISA Conference, Paper #91-0305.

Peterson, W., ''Choosing the Right Temperature Transmitter,'' *InTech,* April 1991.

Plumb, H.H., ''Temperature, Its Measure and Control in Science and Industry,'' 5th Symposium on Temperature held in 1971 by NBS, API, and ISA.

Schooley, J.F., ''State of the Art of Instrumentation for High Temperature Thermometry,'' Argonne National Laboratories Publication Number: ANL-78-7, 1977 Symposium.

Shepard, R.L., et al., ''Development of Long Life, High Reliability, Remotely Operated Johnson Noise Thermometer,'' 1991 ISA Conference, Paper #91-0306.

Stockdale, R.B., ''New Digital Control Systems Need Better Temperature Sensors,'' *Control Engineering,* November 1991.

Tinsley, F.G., and Adams, B., ''Evolution in the Application of Optical Fiber Thermometry,'' 1991 ISA Conference, Paper #91-0307.

Warmoth, D., ''Temperature Measurement & Control,'' *Measurements and Control,* February 1986.

When, S., ''Liquid-In-Glass Thermometers Still Alive and Indicating,'' 1991 ISA Conference, Paper #91-0308.

2 Bimetallic Thermometers

T. J. CLAGGETT, R. W. WORRALL (1969, 1982) **B. G. LIPTÁK** (1993)

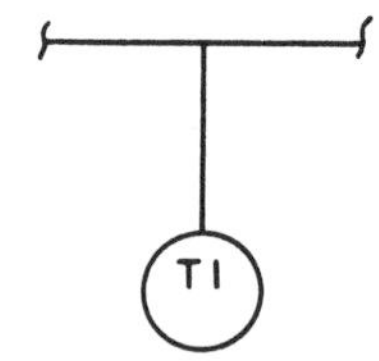

Flow Sheet Symbol

Temperature Ranges:	Minimum standard span is 150°F (50°C); maximum standard span is 800°F (450°C). Standard ranges in approximately 15 steps are available from −100 to 125°F (−70 to 50°C) up to 200 to 1000°F (100 to 550°C)
Materials of Construction:	External parts are usually stainless steel with plastic or glass crystals. The thermowells can be brass, steel, or stainless steel
Dimensions:	Dial diameters vary from 1 to 5 in. (25 to 125 mm); stem lengths vary from 2.5 to 24 in. (63 to 610 mm)
Costs:	A 3 in. diameter, back-connected unit with a 2.5 in. brass thermowell is $75; a 5 in. diameter, multi-angle unit with a 6 in. stainless steel thermowell is $180. Thermometers are usually discounted from 10 to 40% when large quantities are purchased.
Partial List of Suppliers:	Abbeon Cal Inc. (Strap-on); AMETEK Inc. U.S. Gauge Div.; Aschroft-Heise Div. of Dresser Industries; Bacharach Inc.; Bristol Babcock Co. Helicoid Div.; Brooklyn Thermometer Co. Inc.; Cooper Instrument Corp.; Crosby; Dresser Industrial Instrument Div.; Fenwal Inc.; Griffith Industrial Products; Jumo Process Control Inc.; Marshall Instruments Inc.; Marshalltown Instruments Inc.; Marsh Instrument Co.; Moeller Instrument Co.; Noshok Inc.; Pacific Transducer Corp. (Surface); Palmer Instruments Inc.; Princo Instruments Inc.; PSG Industries Inc.; Qualitrol Corp.; Reotemp Instrument Corp.; Stortz, A Div. of PSG Industries Inc.; Taylor Thermometer Corp. of America; Tel-Tru Mfg. Co.; Trend Instruments Inc.; H.C. Trerice; Universal Enterprises; Weiss Instruments Inc.; Weksler Instruments Corp.; Wika Instrument Corp.

Bimetallic thermometers make use of two fundamental principles: (1) metals change volume with temperature and (2) this coefficient of change is not the same for all metals. If two different straight metal strips are bonded together and heated, the resultant strip will tend to bend toward the side of the metal with the lower expansion rate. Deflection is proportional to the square of the length and the temperature change, and inversely proportional to the thickness. A bimetallic spring (Figure 2a) can be calibrated to produce a predictable deflection at a preset temperature. This is the basis of operation of the many bimetallic temperature switches in various household appliances.

The motion produced by a bimetallic spring is small; to amplify it in a reasonably sized space the bimetal strip may be wound in the form of a spiral or a helix. A classical example of an ambient air temperature thermometer is shown in Figure 2b. The outside edge of the spiral is pinned to the frame and a pointer is connected to the center. As the temperature increases the spiral winds up deflecting the pointer clockwise.

Knowing the coefficients of expansion of the two metals, their thickness, and the desired scale length and range, the total length of the spiral can be computed. A favorite combination of metals is low-expanding invar (64% Fe, 36% Ni) against high-expanding nickel-iron alloy with chromium or manganese added.

Most industrial or residential bimetal thermometers use a helical coil which can be designed to fit into a stem more easily than the spiral. The element is surrounded by a protecting tube or well. The device can be mounted to measure the temperature of the gas or liquid inside a duct. The design is frequently used on domestic furnaces and over the years has replaced most of the glass stem thermometers used earlier. Figure 2c illustrates a typical bimetallic thermometer using a helical coil. Also shown is a protective thermowell, which allows the removal or replacement of the thermome-

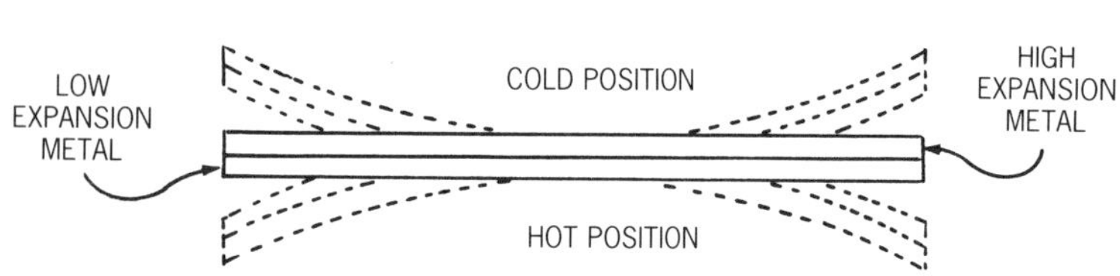

FIG. 2a
The operation of a bimetallic spring.

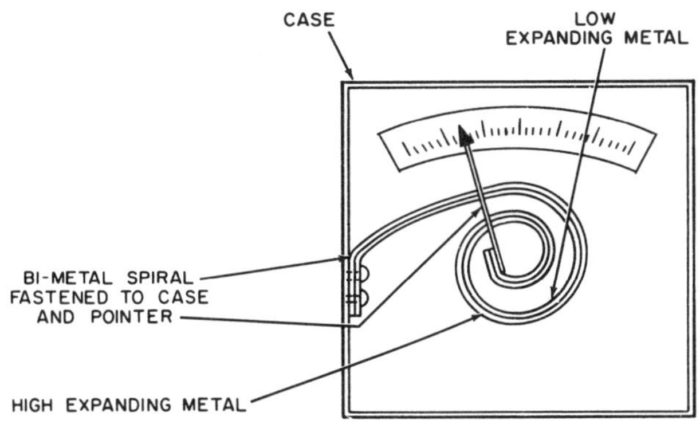

FIG. 2b
Bimetallic ambient air thermometer.

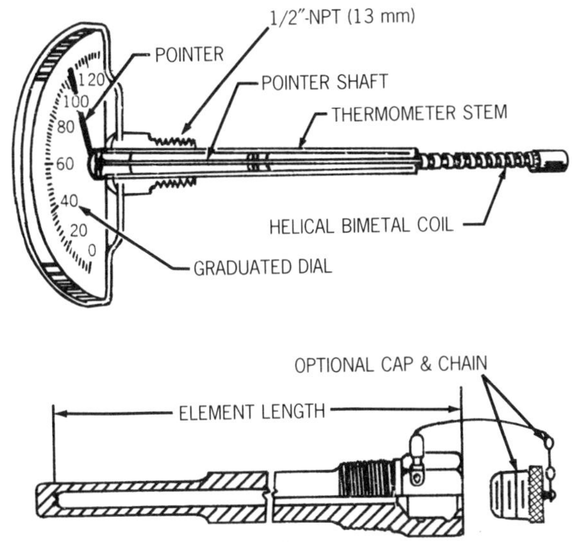

FIG. 2c
Bimetallic thermometer and corresponding thermowell.

ter without opening up the process tank or piping. When only periodic checks are needed on a particular temperature point, a thermowell only is installed and is protected from dirt accumulation by a cap.

A single helix moves axially as it winds or unwinds with heat and cold. This requires clearance for a vertical movement of the pointer. The difficulty can be overcome, if desired, by using a multiple element, wound coaxially so as to form coils within coils. This construction is more costly but has an advantage in requiring less immersion depth (Figure 2d).

The thermometer is usually either back- or bottom-connected, depending on which orientation allows the operator better visibility of the dial face. Bimetal thermometers are also made in types that can adjust the dial face at any angle, with respect to the axis of the stem. This can even be a swivel type as shown on Figure 2e. Any of these constructions requires a bend in the motion transmission from coil to pointer. This is done with an edgewound helical spring, which eliminates backlash and requires little torque to operate.

Another feature usually included is complete sealing. A dry gas is in the dial face portion of the assembly while silicone fluid fills the stem and surrounds the coil to dampen vibration and accelerate heat transfer.

Readout dials are available varying from 1 to 5 in. (25 to 125 mm) in diameter and with stem lengths up to 24 in. (610 mm). Wells, made of carbon steel, stainless steel, or other materials, are available to facilitate removal or to protect against corrosive environments.

Bimetallic elements can be made sufficiently sturdy to actuate a recording pen. A chart, driven by mechanical clockwork behind the pen can form a complete measuring and recording system which is independent of outside electric power.

The advantages over glass stem thermometers include that the bimetallic design is less subject to breakage and is easier to read. The advantages of the bimetallic thermometers relative to the filled or electronic temperature indicators are their lower cost and their simplicity.

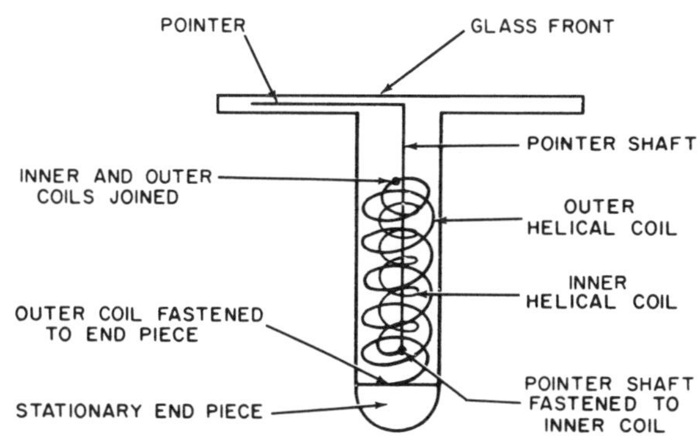

FIG. 2d
Bimetal thermometer with dual helical coil.

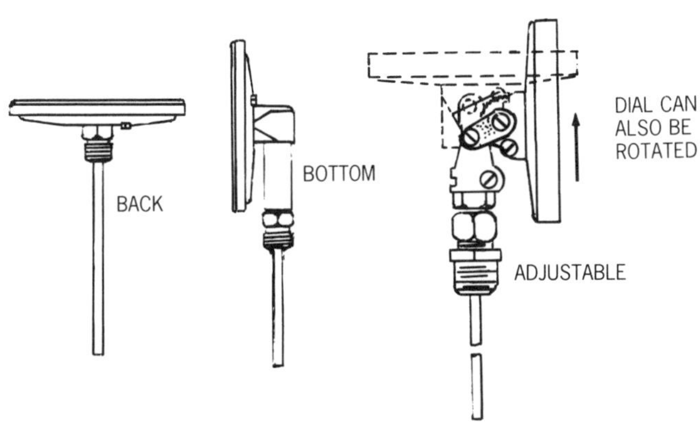

FIG. 2e
The bimetallic thermometers can be back- or bottom-connected, or they can be fully adjustable. (Courtesy of Taylor Thermometer Corp. of America)

A disadvantage is that rough handling changes their calibration. The overall accuracy is not as good as that of the glass stem design. The bimetallic thermometers are confined to local measurement.

Bibliography

''Bimetallic Thermometers,'' *Measurements and Control,* October 1991.

Carlson, D.R., ''Temperature Measurement in Process Control,'' *InTech,* October 1990.

Considine, D.M., *Encyclopedia of Instrumentation and Control,* New York: McGraw-Hill, 1971.

Hashemian, H.M., and Peterson, K.M., ''Assuring Accurate Temperature Measurement,'' *InTech,* October 1989.

Hormuth, G.A., ''Ways to Measure Temperature,'' *Control Engineering,* Reprint No. 948, 1971.

Plumb, H.H., ''Temperature, Its Measure and Control in Science and Industry,'' 5th Symposium on Temperature, held in 1971 by NBS, AIP, and ISA.

Schupp, A.B., ''A Report on Progress in the Field of Bimetallic Actuated Thermometers,'' *Measurement and Control News,* October 1990.

3 Calibrators and Simulators

H. L. DANEMAN (1993)

Types of Designs:	A. Calibrators that use simulators (devices that simulate the output signal of thermometers) B. Calibrators that use calibration baths, which provide accurate temperature environments for calibration purposes (systems use hot air, oil, sand, or aluminum blocks for heat storage; some are fluidized beds)
Costs:	Simple RTD or TC simulator costs start at about \$300. A hand-held thermocouple simulator with 10 memory locations and 0.1% inaccuracy is about \$600; an RTD simulator with 6 decades and 0.005% inaccuracy is about \$2500; a 4 to 20 mA DC transmitter calibrator with simulation of 11 RTDs and 8 thermocouples is about \$6000. Calibration bath costs vary with bath volume, heat storage media, and calibration accuracy. A microprocessor-controlled, programmable bath with IEEE 488 and RS232C interface and 0.002°C stability is about \$12,500; other baths range from \$3000 to \$25,000
Partial List of Suppliers:	Altek Industries Corp. (A); Ametek/Jofra Instruments (B); Automation Service, Test Equipment Div. (B—hot air); Azonix Co.; Beta Products (A); Biddle Instruments (A); Chino Works; Elan Technical Corp. (A); Cole-Parmer (B—fluidized bed); Davis Instrument Mfg. Co. (B); Ever Ready Thermometer Co.; Forma Scientific Co. (B); FTS Systems Inc. (B); Guildline Instruments Inc. (B); Hart Scientific Inc. (A,B); Hotpack Corp. (B); Isothermal Technology Ltd. (B); Joseph Kaye Co.; Loveland Controls Co. (A); Mikron Instrument Co. (A); Neslab Instruments Inc. (B); Omega Engineering Co.; Onicon Inc. (A); Panalarm Div. Ametek Inc. (A); Prime Technology Inc. (A); Procedyne Corp. (B); Promac Inc. (A); Rochester Instrument Systems Inc.; (A); Rosemount Inc., Aerospace Div. (B); Science/Electronics Inc. (B); S-Products Inc. (A); Techne Inc. (A); Tenney Engineering Inc. (B); Thermo Electric Co. Inc. (A); Yokogawa Corp. of America (A)

Temperature calibrators range from simple hand-held instruments to large, permanently installed baths and chambers. Calibrators reproduce temperatures with an accuracy and stability adequate for the range of devices to be checked. The thermometers are normally calibrated at one specific temperature at a time, although a series of baths can be assembled for sequential immersion to enable a range of temperature points to be checked. Correspondingly, the temperature can be reset in stages or as a ramp function for multipoint checking.

Simulators duplicate the outputs of temperature sensors enabling measuring instruments to ''see'' a simulated but precise temperature value. This output may be a millivoltage to simulate thermocouples, a resistance to simulate RTDs, or a light or radiant energy level to calibrate instruments based on optical or infrared (IR) energy. Simulators can usually be connected locally (at the sensor) and then at the readout instrument so as to check the operation of transmitters, multiplexers, and cabling.

A fixed resistor connected to a multipoint monitor is an example of a simulator that continuously checks the calibration of indicators, recorders, and alarms.

Some calibrators incorporate both temperature environment and sensor output functions.

TEMPERATURE CALIBRATION BATHS

A bath or chamber creates a temperature environment suitable for the immersion of temperature sensors such as thermocouples, RTDs, or bulbs. There are two bath types: fixed and adjustable. The earliest fixed-temperature calibrator used was the ice bath. This was supplemented by other so-called fixed points, such as the steam and sulfur points, and later by a full range of freezing, boiling, and triple points. These are based on the principle that materials change state (freeze and boil) at certain fixed temperatures (Table 3a). Many of these points have become reference values for defining the temperature scale and are therefore especially appropriate for calibration.

Useful examples of such fixed points are the triple point of water (0.0100°C) and the melting point of zinc (420°C). The triple point of water is obtained in a sealed container in which the solid, liquid, and vapor states of water are in equilibrium (Figure 3b). In contrast to the triple point, the ice point can be more easily obtained and used to an accuracy suitable for industrial calibration. RTDs in particular are defined in terms of their R_0 value, or resistance value at the ice point (e.g., 100Ω Pt.). Immersion in an ice point bath is a useful way of compensating for calibration drift even though the thermometer is normally used at higher temperatures.

TABLE 3a

Defining Fixed Points of the International Practical Temperature Scale (IPTS-90)

Point Number	Temperature $T_{90}/°K$	Temperature $t_{90}/°C$	Substance	State
1	3 to 5	−270.15 to −268.15	He	VP
2	13.8..	−259.35..	eq. H_2	TP
3	≈17	≈−256.15	eq. H_2 (or He)	VP (or GT)
4	≈20.3	≈−252.85	eq. H_2 (or He)	VP (or GT)
5	24.6..	−248.5..	Ne	TP
6	54.5..	−219. . .	O_2	TP
7	83.8..	−189. . .	Ar	TP
8	234. . .	−39. . .	Hg	TP
9	273.16	0.01	H_2O	TP
10	303. . .	29.77..	Ga	MP
11	430. . .	157. . .	In	FP
12	505. . .	231. . .	Sn	FP
13	693. . .	420. . .	Zn	FP
14	934. . .	660. . .	Al	FP
15	1235. . .	962. . .	Ag	FP
16	1338. . .	1064. . .	Au	FP
17	1358. . .	1085. . .	Cu	FP

FP = freezing point
MP = melting point
GT = gas thermometer point
TP = triple point
VP = vapor pressure point
eq. H_2 = hydrogen at the equilibrium concentration of the ortho- and paramolecular forms

Adjustable temperature baths contain a fluid (liquid or fluidized solid) circulated through a chamber in which the thermometers can be immersed. Controllers maintain a temperature at the desired setpoint. A block establishes temperature uniformity amongst sensors at temperatures below 1472°F (800°C), or where conduction is the primary means of heat transfer.

SIMULATORS

Simulators for thermocouples and RTDs consist of voltage sources and/or resistors having values that correspond to the required temperature readings. Thermocouple simulators require provision for reference junction compensation and alloy terminals.

Simulators for optical or total radiation pyrometer calibration may be as simple as a 25 watt lamp or as sophisticated as a controlled output arc furnace. It is usual to use these simulators with comparison standards. By comparing one optical pyrometer with a standard or with a laboratory-calibrated optical pyrometer one can avoid errors that may arise due to their large ambient temperature coefficients. Such calibrations can, for example, be performed in an outdoor shed or pyrometer shop adjacent to a steel mill.

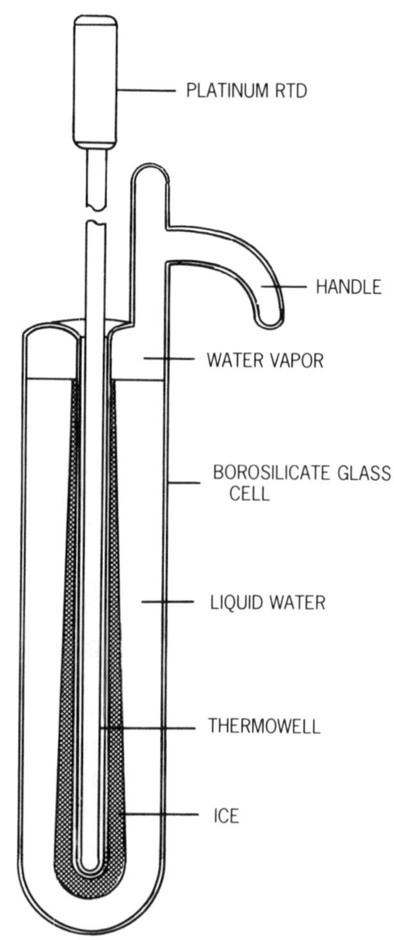

FIG. 3b
Sealed container used to obtain the triple-point temperature of water used in the calibration of RTDs.

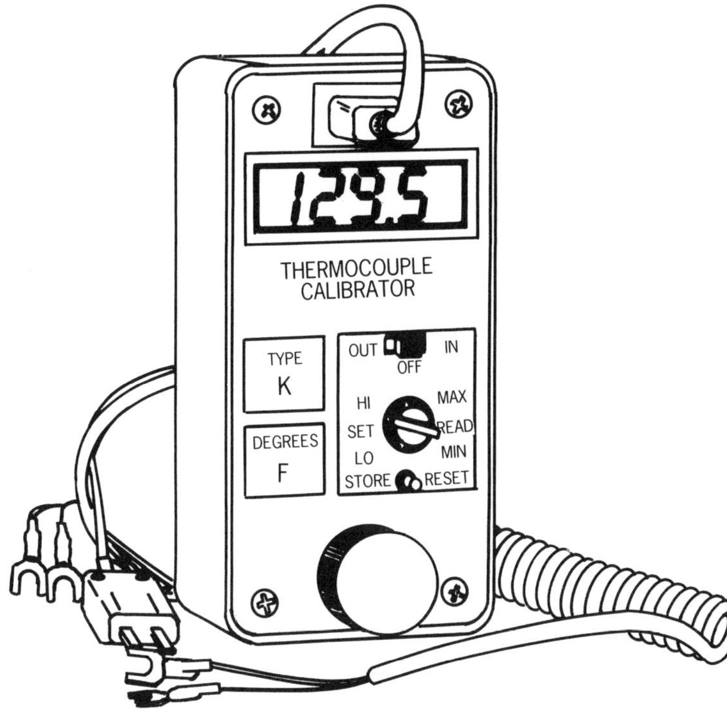

FIG. 3c
Hand-held thermocouple calibrator. (Courtesy of Altek Industries Corp.)

A typical hand-held calibrator is illustrated in Figure 3c. A laboratory-type thermocouple calibrator/simulator is shown in Figure 3d. This unit combines both the simulation and calibration functions in one unit.

A simulator for optical pyrometry is shown in Figure 3e. The unit incorporates a copper fixed point to provide a reference temperature 1985°F (1085°C) independent of a standards laboratory. A very high temperature 3002°F (1650°C) thermocouple and optical/radiation pyrometer calibration system incorporating a palladium freezing point,

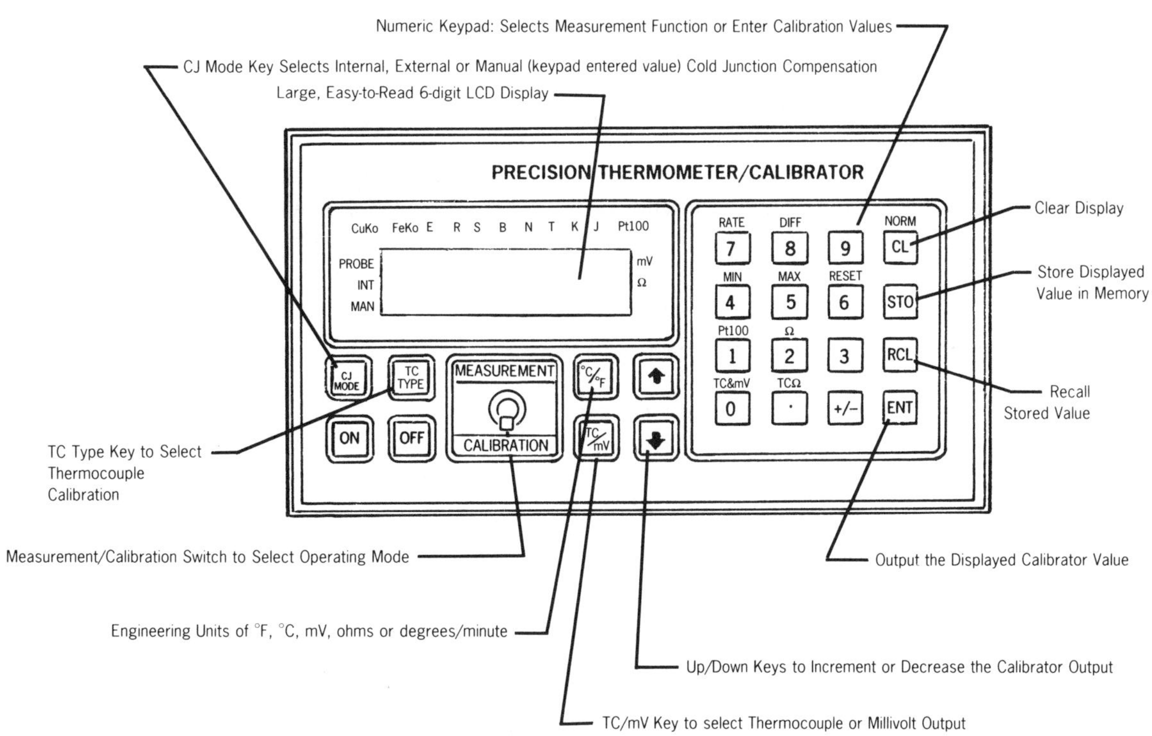

FIG. 3d
Thermocouple calibrator/simulator used in the laboratory.

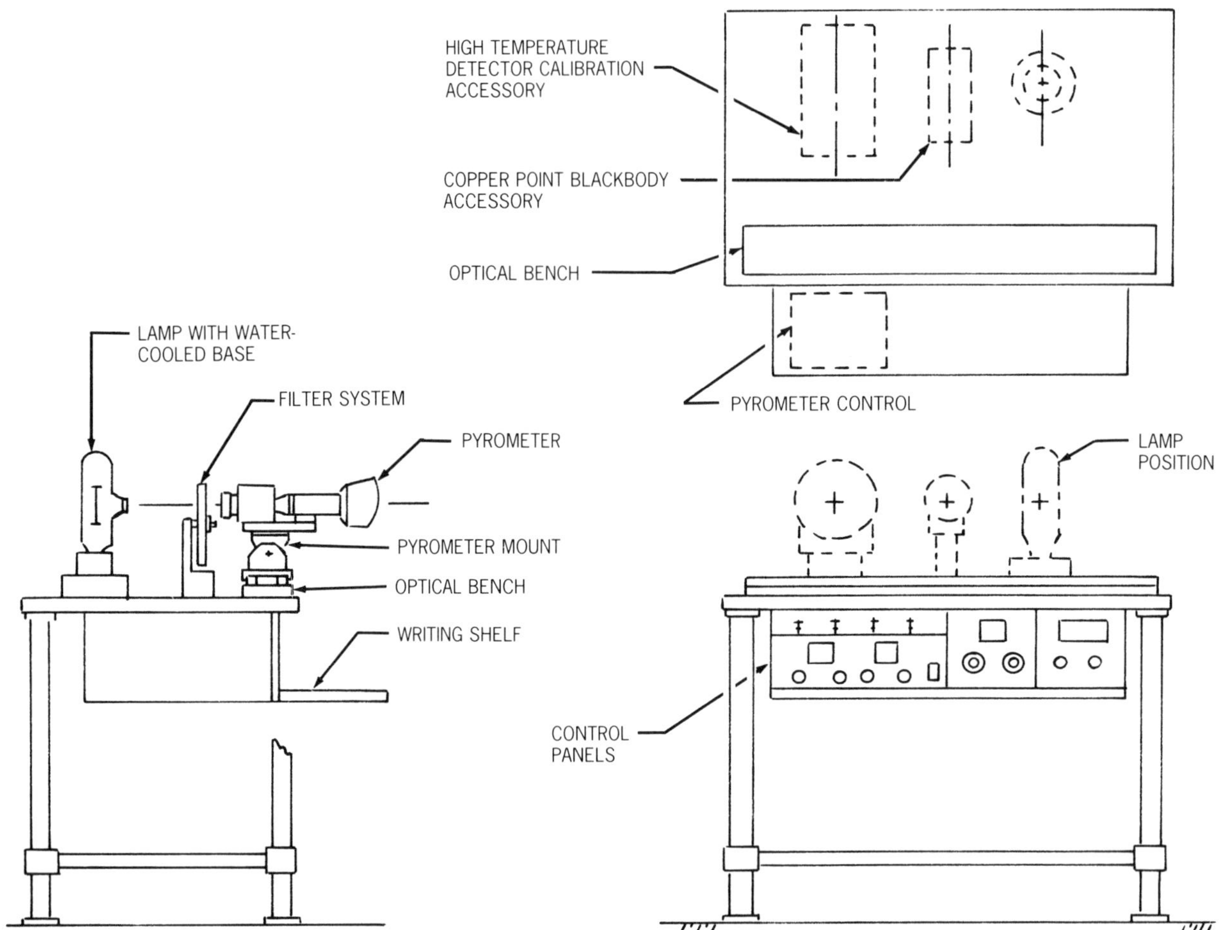

FIG. 3e
Two pyrometer calibration system with copper point, blackbody furnace, and standard lamp.

molybdenum block, and standard lamp was built for a steel mill in Brazil where traceability to a national laboratory was unavailable.

Field and laboratory temperature calibrations are best performed with calibrators and simulators that are adapted for the particular application.

CONCLUSIONS

Table 3f provides some recommendations for the type of calibration equipment that is best suited for particular temperature sensors and industrial or laboratory environments.

TABLE 3f
Calibration Equipment Recommended as a Function of the Type of Thermometer and of the Operating Environment

Temperature Sensor Being Calibrated	*Laboratory Environment*	*Industrial Installation*
Thermocouples	Triple Point	Automatic Ice Point
	Metal Freezing Points	Portable Calibrator
	Fluidized Bath	Portable Simulator
RTDs	Triple Point	Portable Calibrator
	Freezing Points	Portable Bath
	Laboratory Simulator	Portable Simulator
Optical Pyrometers	Comparison System	Lamp Comparator
Radiation Detectors	Comparison System	Optical Pyrometer

Bibliography

ASTM, *Annual Book of ASTM Standards,* Vol. 14.

"Calibration Technology Goes Digital," *InTech,* September 1989, pp. 86–87.

Daneman, H.L., "The Do's and Don'ts of Temperature Calibration," NCSL Annual Conference Proceedings, June 1991.

Daneman, H.L., "The Precise Calibration of Industrial Pyrometers to 1,650°C," Measurement Science Conference, Anaheim, California, January 1986.

Gray, W.T., "Calibration of Optical Pyrometers," ISA Transactions, Vol. 6, 1967.

Kaufman, A., "Liquid Cooling of Electronics," *Measurements and Control,* September 1979, pp. 120–121.

Kaufman, A., "Temperature Transducer Calibration Baths," *Measurements and Control,* February 1986, pp. 188–189.

Kaufman, A., and Drees, W., "How Accurate Are Your Transducer Calibrations?" *Instruments & Control Systems,* November 1959, pp. 1682–1685.

Kaufman, A., and Mitchell, P., "How Accurate Are Your Temperature Reference Baths?" *Instruments & Automation,* March 1955, pp. 450–451.

Mangum, B.W., and Furukawa, G.T., "Guidelines for Realizing the International Temperature Scale of 1990 (ITS-90)," NIST Tech. Note 1265, August 1990.

Michalski, L., Eckersdorf, K., and McGhee, J., *Temperature Measurement,* John Wiley & Sons, 1991.

NBS Monograph 174—Thermometer Calibration.

"Standards, Simulators, Calibrators," *Measurements and Control,* September 1991.

"Temperature Calibration Baths," *Measurements and Control,* September 1991.

Withers, P., "Thermocouple Calibration," *Measurements and Control,* September 1990, pp. 144–147.

4 Color Indicators, Crayons, Pellets

T. J. CLAGGETT, R. W. WORRALL (1969, 1982) **B. G. LIPTÁK** (1993)

Temperature Settings:	Crayons and paints are usually rated with a combination of temperature and time, meaning that a color change is expected to occur if a particular temperature is held for a particular time period
Temperature Ranges:	Paints and crayons are available for every few degrees from 100 to 2500°F (40 to 1371°C)
Costs:	\$3 per ounce of paint or \$5 per crayon; skin temperature detector strips are 5 cents each
Inaccuracy:	Indication is affected by speed of heat-up and by time spent at target temperature—usual error is 10 to 20°F (5 to 10°C). The inaccuracy of most crayons is 1% of the rating. Liquid crystal strips can detect skin temperature within 1°F (0.5°C)
Partial List of Suppliers:	Ball Research Inc. (cone); Biosynergy Inc. (freeze-thaw indicator); Electronic Development Labs Inc. (crayons); Edward Orton Jr. Ceramic Foundation (cones); Hub Material Co. (crayons); Markal Co. (stick-ons); Shell Oil Co.; Solder Absorbing Technology Inc. (crayons, laquers); Tempil Division of Big Three Industries Inc. (crayons, paints); Thermax USA Div. of Everest Assoc. (paints, crayons, labels)

A number of temperature-related physical changes have been used to produce simple thermometers. Crayon marks on heated work-pieces or pellets placed in furnaces change from solid to liquid when their melting point is reached; paints and heat-sensitive labels change their color; luminescent materials change their brightness; and liquid cholesteric crystals detect skin temperature. (Pyrometric cones are discussed in Section 10; the color indicators are covered here.)

For centuries, manually operated furnaces have been temperature-controlled by the operator's placing a heat-sensitive object inside the furnace and observing the status of that indicator through a peephole. The most easily observed physical changes were found to be melting and changes in color. Colored crayons, paints, and pyrometric cones have been used as temperature indicators and are still used as end-point indicators in a number of high-temperature batch processes.

Color indicators are a class of sensors which have the property of changing their original color when a certain temperature is reached. The change is distinct, not just an alteration in shade. For instance, an indicator may change from yellow to grey, or from light blue to light brown. Some can go through several color changes at different temperatures.

Paints and crayons are familiar forms of these indicators which are applied directly to a solid object either when it is cold and about to be heated or when it is already hot. Some indicators can determine the temperature of solid objects immersed in oil. They are not recommended for use in hot gases.

Temperature is indicated by a chemical reaction, where a molecule of a gas, such as ammonia, CO_2, or water vapor, is driven off the basic stock (colorful salts of metals like nickel, cobalt or chromium), thus changing its color. The change is usually permanent after the object cools down. An exception occurs when the gas is water vapor; the indicator may slowly reabsorb this gas from the air and revert to the original color.

Change in color of these types of indicators is not only a function of temperature but also of time. For this reason, the immediate past temperature history of the indicator will influence the exact point at which it will change color. The indicators are usually rated for a specific temperature over a certain time period, for instance 140°F (60°C) in 30 minutes. This means that if held at a constant 140°F the color change will occur in 30 minutes. If the color change occurs in less than 30 minutes, the average temperature is higher than 140°F, and vice versa. On such an indicator, if the temperature does not exceed 130°F (54°C) the change will never occur, because the indicator is stable below this temperature.

Figure 4a shows typical time-temperature relationships for two different crayons. In the examples shown, the temperature at which color change occurs is quite critical when exposure time is short. For lower soakout temperatures, the changeover will occur in a longer time.

Flatter curves than those shown are possible where changeover will occur within a few seconds after "operating" temperature is reached, or it won't occur at all.

Many different temperature ratings are available. They can be obtained in a series for every few degrees to the maximum offered (about 2500°F, or 1371°C). This class of indicators is quite inexpensive and is used in industry where

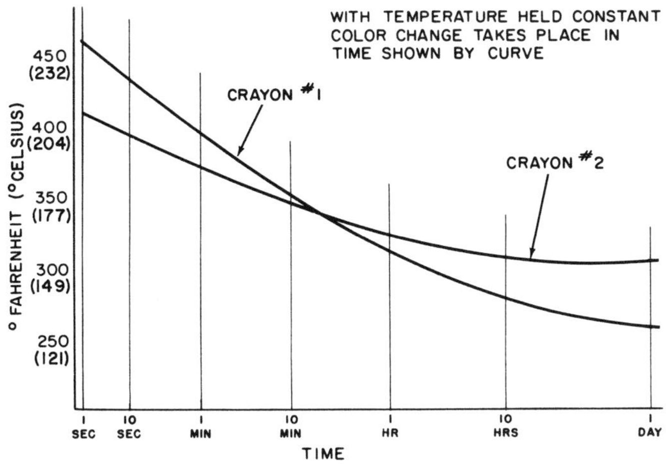

FIG. 4a
Time-temperature relationship for color indicators.

only an end point is needed and someone can be present to watch for or interpret the results. A disadvantage of these sensors is that the material adheres tightly to the object on which it is placed and presents a problem if it must be removed later.

Bibliography

Considine, D.M., *Encyclopedia of Instrumentation and Control,* New York: McGraw-Hill, 1971.

Hormuth, G.A., ''Ways to Measure Temperature,'' *Control Engineering,* Reprint No. 948, 1971.

Plumb, H.H., ''Temperature, Its Measure and Control in Science and Industry,'' 5th Symposium on Temperature, held in 1971 by NBS, AIP, and ISA.

''Temperature-Sensitive Paints, Crayons,'' *Measurements and Control,* December 1991.

5 Fiber-Optic Thermometers

H. L. DANEMAN (1993)

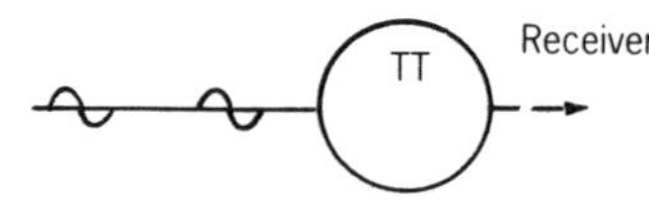

Flow Sheet Symbol

Types of Sensors:	Light-pipe, blackbody, dual-wavelength, crystal, gap, and fluoroptic
System Components:	Fiber-optic assembly, infrared detector, and electronic console
Applications:	Temperature measurement of hard-to-reach objects in, for example, calendering, crystal growing, engine heads (spark plugs), glass forehearths, induction heating, kiln hot spots, medical hyperthermia, microwave packaging, polymer melting, printer operations, reactive-ion etching, transformer windings, vacuum processing, web drying, and welding
Wavelengths Used:	0.7 to 8 microns for inexpensive total radiation detectors, 2.2 to 3.8 microns for penetrating intervening atmosphere, 2.2 microns for metals, 4.5 microns for flames, under 8 microns for glass
Field of View:	The ratio of target distance to target size can range from 3:1 to 900:1
Fibers:	Single fiber is 0.5 mm in diameter, 4-fiber array is 0.9 mm in diameter. The number of fibers per cable ranges from 30 to 400; their lengths range from 10 to 300 ft (3 to 100 m)
Temperature Ranges:	Can detect from 212°F (100°C), but the typical range is from 500 to 5500°F (260 to 3000°C); some units go up to 6500°F (3600°C). Allowable ambient temperature range is from −76 to 535°F (−60 to 315°C)
Transmission Range:	0.7 to 8 microns
Spectral Response:	2.0 to 2.4 microns
Response Time:	0.3 seconds or better
Total Acceptance Angle:	22 to 60 degrees
Stability:	Drift is under 0.5°F (0.3°C) over period of 10 days
Inaccuracy:	From as low as 0.2°F (0.1°C) to 1% full scale
Costs:	Transmitters are in the \$2000 to \$4000 range. A 0.1°F resolution microprocessor-based unit with PID algorithm included is about \$6000; the same unit with 2, 4, 6, or 8 channels is about \$14,000
Partial List of Suppliers:	Accufiber Div. of Luxtron Corp.; ASEA Innovation; Barber-Colman Co.; Hart Scientific; Ircon; Irtronics Instruments Inc.; Land Instruments Inc.; MetriCor Inc.; Mikron Instrument Co. Inc.; Square D, Infrared Measurement Div.; Technology Dynamics Inc.; Vanzetti Systems Inc.; Williamson Corp.

Radiation pyrometry is discussed in Section 11. This section will concentrate on combining those principles with the use of optical fibers, which provide the ability to look at the temperature of small objects or to ''look around'' opaque objects.

Noncontact thermometry has many advantages, particularly in regard to measuring temperatures of objects that are extremely hot or corrosive; are moving or fragile; are in strong electromagnetic (RF, microwave, or DC magnetic) fields; or are subject to measurement error due to heat loss by conduction. Until the glass-fiber optic cable became available, noncontact thermometry required line-of-sight vision between the sensor and the target object. The elimination of this restriction makes it possible for the fiber-optic thermometer to solve many difficult measuring problems.

Optical fiber thermometry (OFT) depends on total internal reflection within a thin fiber element. Absorption of infrared energy by glass fibers has limited the low end of the range to 660°F (350°C) (see Figure 5a). Developments in telecommunications and glass research now promise to facilitate lower temperature measurements using alternative spectral bands (see Table 5b).

Some laboratory-standard fiber-optic instruments offer even greater accuracy than a type S thermocouple—±0.01% at 1832°F (1000°C) with a resolution of 0.01°C.

THERMOMETER DESIGN

Fiber-optic measurement systems consist of three elements: the fiber-optic assembly, an infrared detector, and an electric console. A single fiber or bundle of several fibers gather infrared radiation from the target, transmit to the detector, and convert the radiation to a voltage suitable to the required function.

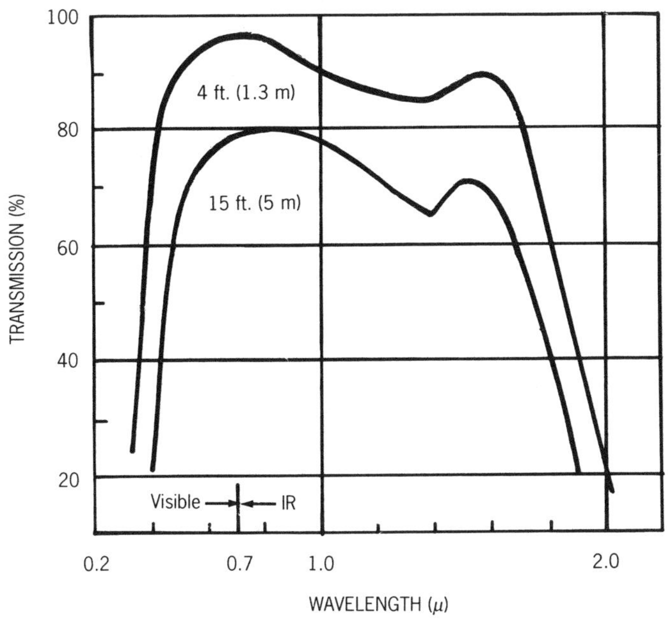

FIG. 5a
The infrared transmission of glass fibers designed for high-temperature applications varies according to cable length. The sensors used for this purpose are filtered in the 0.8 μm region.[1]

Fibers are sensitive only to the infrared portion of the spectrum and filter out other radiation; thus they are not activated by flames and fumes. Even though infrared radiation becomes detectable at about 140°F (60°C), it is best to use 212°F (100°C) as the starting point in any monitoring application.

Fibers, whether single or in bundles, are always enclosed in metal or ceramic sheaths for protection. A unique characteristic of the special glass fibers is the ability to bend light and to transmit it for distances up to 30 ft (9.1 m) without distortion or loss of definition.

Perhaps the one outstanding feature of fiber optic systems is the ability of the fibers to withstand and function in hostile environments, including intense heat. Fibers maintain resolution exceeding one degree change at temperatures over 2000°F (1100°C). At extreme high temperatures and under other severe conditions, the fibers are protected with air or inert gas purging or by water cooling.

The advantages of using fiber optics in noncontact temperature measurement applications include their inertness, relatively rugged design, small size, and ability to ''look around'' opaque objects (Figure 5c). With the addition of a telescopic lens system, the same fiber-optic assembly can be made to monitor different target areas at various focal distances.

Radiation proportional to temperature is conducted along an optical fiber to a sensor connected to a measuring instrument. Variations on this basic principle make it possible to adapt this design to a variety of applications.

As with any optical instrument, a fiber-optic thermometer depends on observations of brightness or on total radiation. The measurement technique most often used is characterized by the distribution of energy by wavelength and field of view.

A number of factors affect the choice of wavelength. For example, general-purpose instruments use a wide energy band (8 to 20 microns). Inexpensive instruments cover 0.7 to 8 microns in order to include most of the total energy radiated by the hot object. For penetration of intervening atmosphere, shorter wavelengths of 2.2 to 3.8 microns are preferred. For metals (which have a high reflectance), the narrow band around 2.2 microns reduces the effect of emissivity variations. (For definitions of terms such as ''emissivity,'' see Section 11.) Glass is more transparent at the shorter wavelengths (for example, 0.8 microns). Reflection becomes critical above 8 microns, so the shorter wavelengths are used for molten glass temperature measurement. Flames (preferably clean) can best be measured at a wavelength of 4.5 microns.

Field of view is the other significant parameter characterizing optical measuring instruments. This is described in terms of ratio of the target distance to the target size, and ranges from 900:1 to 3:1. A single wavelength instrument measures the average temperature of the body filling its field of view. A dual-wavelength instrument measures the hottest part of the target within its field of view.

TABLE 5b
Infrared Fiber-Optic Cable Comparative Specifications

	Glass	*Quartz*	*IR*
Fiber Diameter	0.002 in.	0.008 in.	0.008 in.
Fibers/Cable	200–400 fibers/bundle	50 fibers/bundle	30 fibers/bundle
Cable Diameter	0.063 in.	0.063 in.	0.063 in.
Transmission Range	0.5 to 1.4 μm	0.7 to 2.4 μm	0.5 to 4.5 μm
Ambient Limits	500°F	500°F	300°F
Total Acceptance Angle	68°	50°	22°
Sensor Temperature Limits	1000–5000°F	200–2000°F	125–2000°F
Flexibility Rating	excellent	good	limited

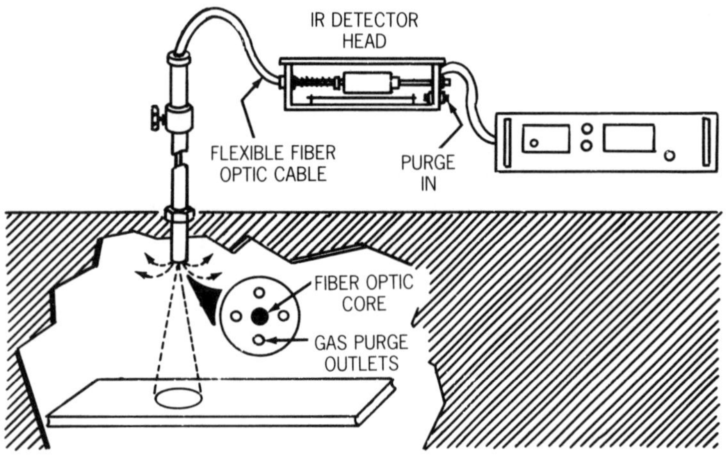

FIG. 5c
The optical fiber thermometer consists of three components: the OFT assembly, the infrared detector, and the electronic console. (Courtesy of Vanzetti Systems Inc.)

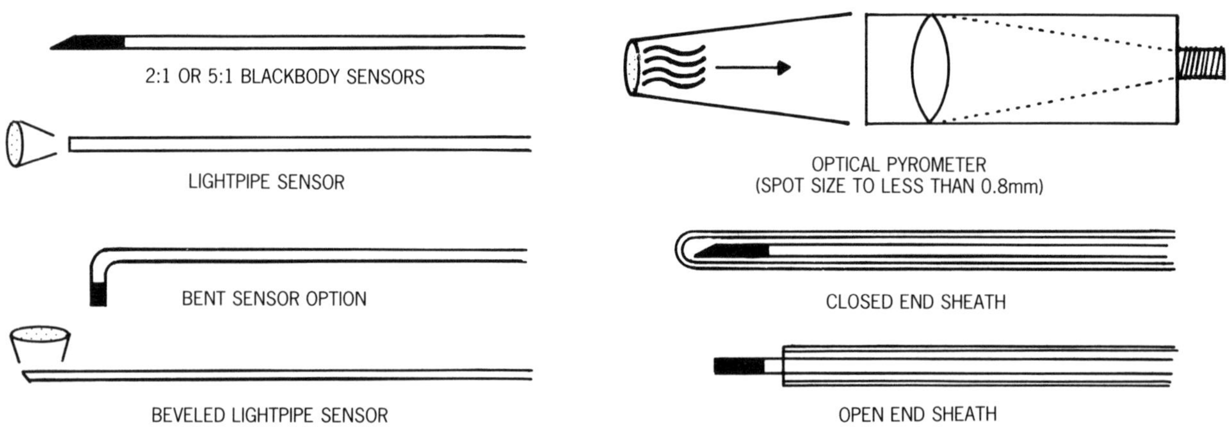

FIG. 5d
The sensor portion of the OFT probe can be configured in many ways.

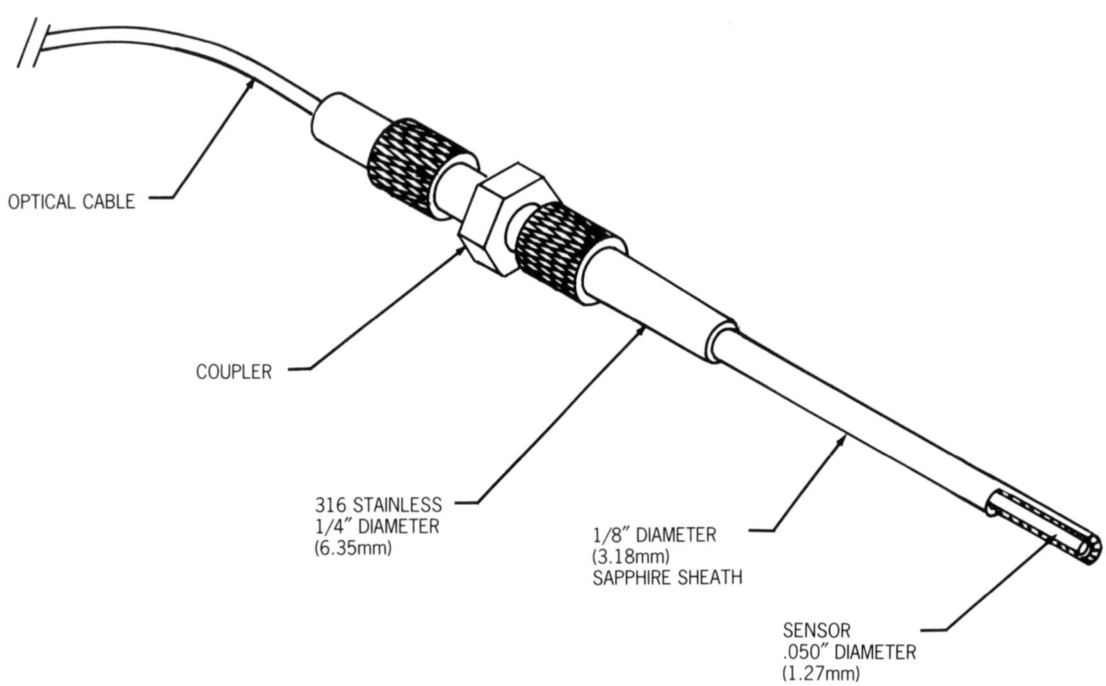

FIG. 5e
Open-ended sheath design of the light-pipe sensor. (Courtesy of Accufiber Div. of Luxtron Corp.)

THE OPTICAL FIBER THERMOMETER (OFT) SENSORS

The available variations in fiber-optic probe sensor designs are shown in Figure 5d. The fiber-optic sensors fall into four categories: light-pipe, blackbody, dual-wavelength, and gap types.

The light-pipe design transmits the radiation from the target to the detector through an open tip, as shown in Figure 5e. The blackbody-type unit radiates heat from a cup of material, such as a thin coat of precious metal, surrounding the tip.

A sapphire optical rod is coated at the tip with a thin film of irridium or platinum and aluminum oxide. At elevated temperatures, the tip emits a band of wavelengths which are transmitted to a detector for measurement (Patent No. 4,576,486). A schematic of the complete system is shown in Figure 5f.

In the dual-wavelength optical pyrometer sensor the radiation is focused through a lens onto the optical fiber link from a distance of up to 16 ft (see Figure 5g).

In the design shown in Figure 5h, light of a fixed wavelength spectrum is transmitted from the instrument through the optical fiber to the sensor. Here the light is converted and returned as a different wavelength spectrum, the value of which *depends on the temperature* of the sensor. The *wavelength* of the *returned* light is scanned and presented as a

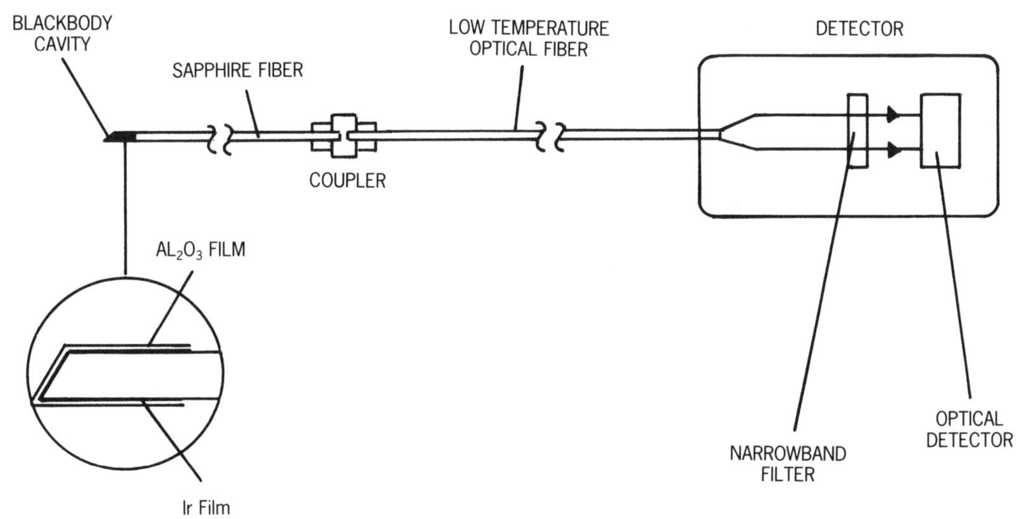

FIG. 5f
Blackbody-type OFT sensor system. (Courtesy of Accufiber Div. of Luxtron Corp.)

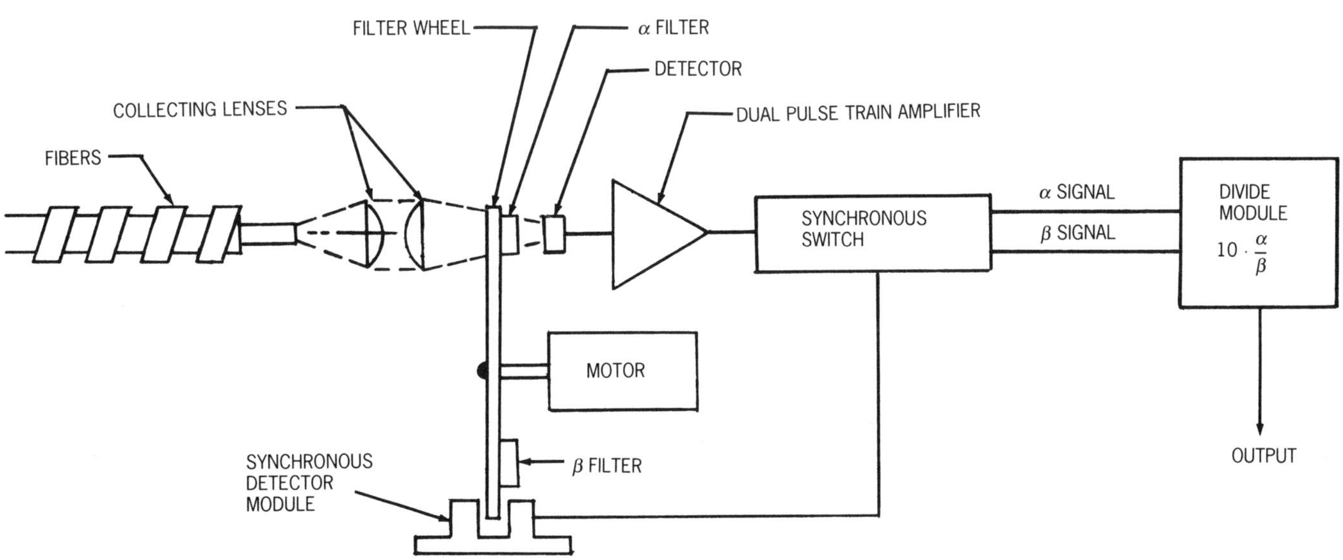

FIG. 5g
This dual-wavelength design measures radiant flux at two separate wave bands and calculates the signal ratio to determine target temperature. The technique can measure low and varying emissivity targets and small moving objects that do not fill the field of view.[1]

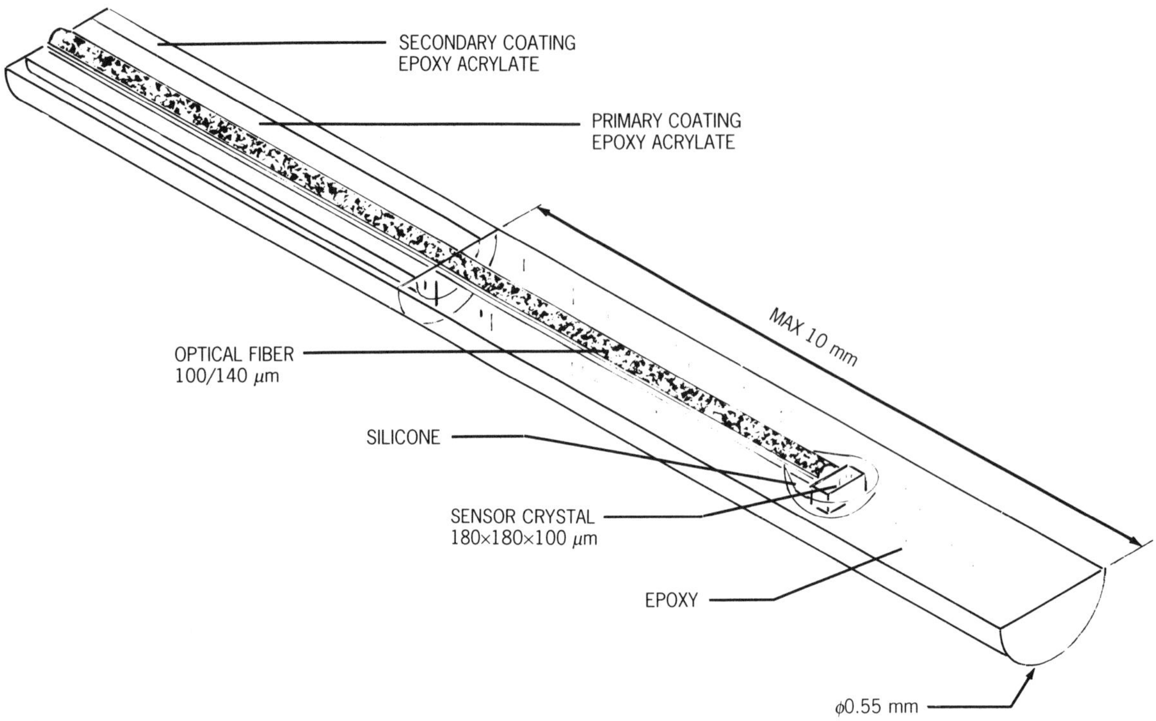

FIG. 5h
Crystal-type miniature fiber-optic thermometer. (Courtesy of ASEA Innovation)

temperature reading on the instrument display or output to one of the optional interfaces.

The Fabry-Perot thermometer uses a temperature-sensitive spacer which varies a gap width, thereby establishing a series of spectral bands by means of optical interference (Figure 5i). Because it is entirely optical from the measurement end of the fiber to the detector, it can be made very small. The measuring range is 35 to 55°C with a resolution of 0.1°C. The detector can be a pyroelectric detector amplified by an FET preamplifier.

The fluoroptic sensor measures the decay time of a fluorescent material (magnesium fluorogermanate) which, after

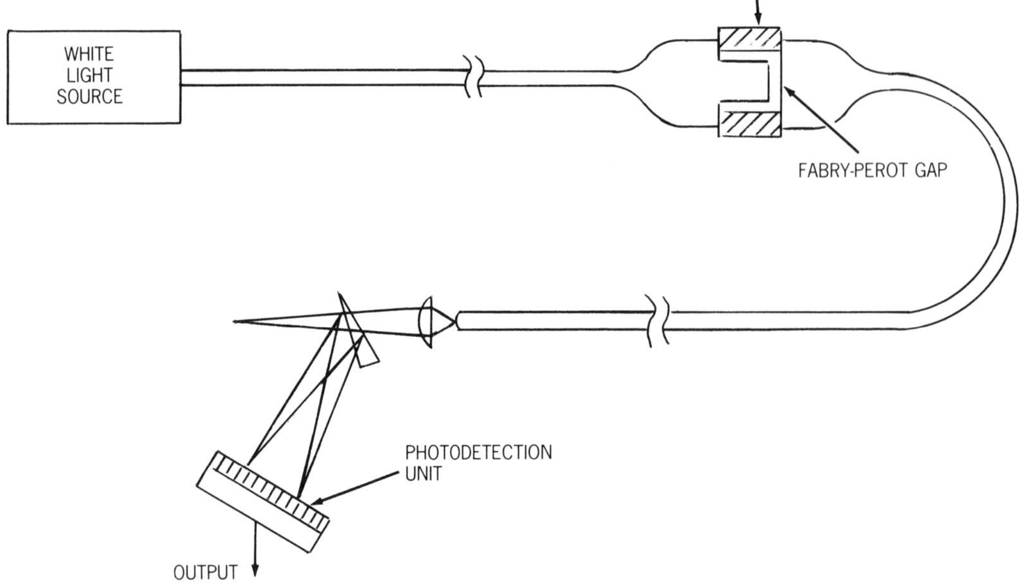

FIG. 5i
Fabry-Perot gap-measuring system.

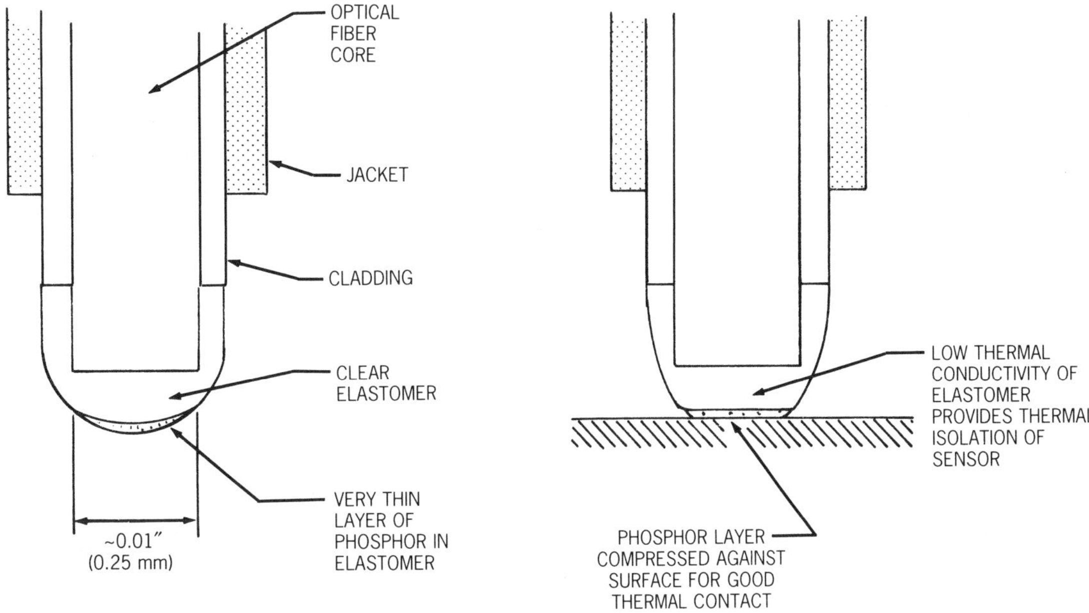

FIG. 5j

In this contact probe, the phosphor sensing layer is on the outer surface of a transparent elastomeric tip that conforms to the surface of interest.[2]

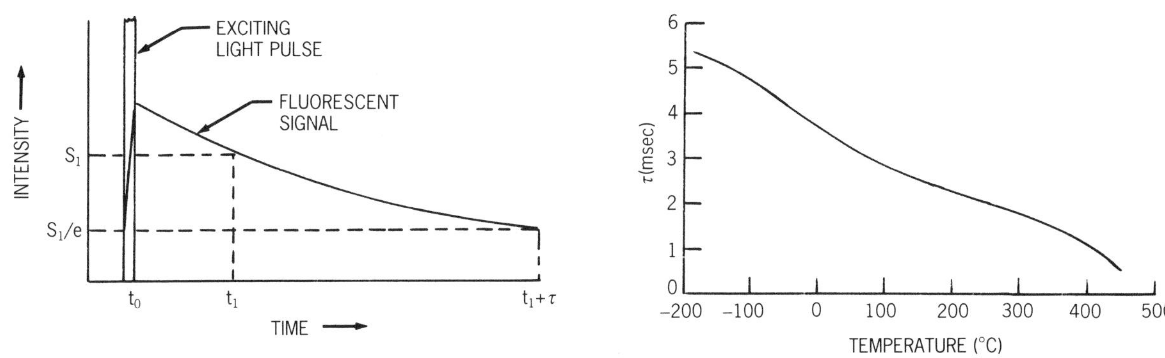

FIG. 5k

Decay and calibration characteristics of the phosphor sensor.[2] *Left: Method for measuring the fluorescent decay time of the phosphor sensor. The time between the initial measurement of the signal level,* S_1*, and the crossover of the decaying signal with* $S_{1/e}$ *is the decay time. Right: Plot of decay time versus temperature for the magnesium fluorogermanate phosphor sensor. This is the basic calibration curve for the system.*

being energized by a short-wavelength light pulse, varies proportionally with temperature. The phosphors can be compressed into intimate contact with a surface to measure with little conduction loss (Figure 5j).

In Figure 5k the fluorescent signal decay characteristics of the fluoroptic sensor are shown on the left, and the basic calibration curve of the phosphor sensor is shown on the right.

CONCLUSIONS

The advantages of OFT pyrometry include the small size of the sensor, which does not require line-of-sight observation of the object and can be furnished with blackbody fibers. OFTs are not affected by RF, microwave, or electromagnetic fields or by shock and vibration; their range is wide; their response is fast; and they can average or provide temperature profiles through noninvasive remote measurements of temperatures of solid objects or immersed in liquids. The main disadvantage of OFT pyrometry is the high unit cost of this measurement. The unit cost can be reduced through multiplexing, so that several OFT sensors are connected (through a multiplexer) to a common set of electronics.

Fiber-optic thermometers offer many options resulting in application to a wide range of industrial temperature measurements. A principal advantage is electrical and, sometimes, even physical isolation from the target. The glass or quartz fibers which transmit the temperature signal to a detector can be an integral part of the measurement reducing lag and conduction error. This is true in the case of the blackbody sensor which incorporates a blackbody cavity on the surface of the fiber tip and the surface temperature sensor adhering to a phosphor emitter pressed onto its tip.

When selecting a fiber-optic thermometer for a given application, it is necessary to review all types of fiberoptic instruments for wavelength and field of view in order to optimize performance.

References

1. Barron, W.R., ''Fiberoptic Infrared Thermometry,'' *Sensors,* September 1990, pp. 74–77.
2. Wickersheim, K.A., and Sun, M.H., ''Fiberoptic Thermometry and Its Applications,'' *Journal of Microwave Power,* 1987, pp. 85–94.

Bibliography

Adrian, P., ''Technical Advances in Fiberoptic Sensors: Theory and Applications,'' September 1991, pp. 23–45.

de Santoli, L., Ruffino, G., and Santoboni, S., ''Calibration of a Fiberoptics Radiation Thermometer,'' *High Temperature High Pressure,* Vol. 5, 1973.

Dils, R.R., US Patent 4,576,486.

''Fiberoptic Sensing of Physical Parameters,'' *Sensors,* October 1988, pp. 21–26.

''Fiberoptic Thermometry for Difficult Surface Temperature Measurements,'' *Sensors,* September 1990, pp. 17–21.

Katzir, A., Bowman, H.F., Asfour, Y., Zur, A., and Valeri, C.R., ''Infrared Fibers for Radiometer Thermometry in Hypothermia and Hypothermia Treatment,'' IEEE Transactions on Biomedical Engineering, Vol. 36, June 1989, pp. 634–637.

Krigman, A., ''Guide to Selecting Non-Contact Temperature Instrumentation,'' *InTech,* June 1983, pp. 23–40.

Ovren, C., Adolfsson, M., and Hok, B., ''Fiber-Optic Systems for Temperature and Vibration Measurements in Industrial Applications,'' Conference on Optical Techniques in Process Control, The Hague, June 14–16, 1983.

Pepe, R.C., ''Fiberoptic Calibration Considerations,'' NCSL, July 1987, pp. 66-1 to 66-12.

''Radiation Thermometry,'' *Measurements and Control,* February 1991, pp. 192–206.

Saaski, E.W., Hartl, J.C., and Mitchell, G.L., ''A Fiberoptic Sensing System Based on Spectral Modulation,'' ISA, 7/1986/1177-1184.

Stockdale, R.B., ''New Digital Control Systems Need Better Temperature Sensors,'' *Control Engineering,* November 1991.

Wickersheim, K., and Sun, M., ''Phosphors and Fiberoptics Remove Doubt from Difficult Temperature Measurements,'' *Research and Development,* Vol. 14, November 1986.

Wohlstein, S., ''Fiberoptics,'' *Measurements and Control,* September 1991.

6 Filled-Bulb and Glass-Stem Thermometers

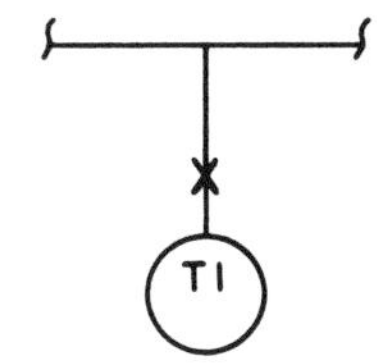

Flow Sheet Symbol

T. J. CLAGGETT, R. W. WORRALL (1969, 1982) **B. G. LIPTÁK** (1993)

Applications:	Glass-stem units are used only as thermometers; filled elements are also used as switches, transmitters, and controllers
Designs:	Glass-stem units are straight, back-angle, and multiangle thermometers; filled systems are frequently surface- or flush-mounted and connected to the process by capillary
Materials of Construction:	Industrial glass-stem thermometers have aluminum, Cycolac, or other plastic cases; spring-loaded glass or plastic windows; and stainless steel stems. For the filled systems, the cases can be aluminum, brass, phenol, polypropylene, nickel-plated steel, or stainless steel; the thermometer movement is usually either bronze or stainless steel; and the bulb and the capillary can be copper, steel, Monel, Hastelloy, or stainless steel. Thermowells can be brass, bronze, steel, stainless steel, Monel, or Hastelloy
Scales and Dials:	Glass-stem thermometer scales are usually straight, with 5, 7, or 9 in. (125, 175, or 225 mm) lengths. Filled thermometer scales are usually circular and have 3.5, 4.5, 6, 8.5, or 12 in. (88, 113, 150, 213, or 300 mm) diameters
Bulb Sizes:	Mercury-, vapor-, and gas-filled bulbs are usually $^{11}/_{16}$ in. diameter by 4 in. (17 by 100 mm), while liquid-filled bulbs are usually $^{3}/_{8}$ by 5 in. (9 by 125 mm)
Capillary Lengths:	Standard lengths from 5 to 80 ft (1.5 to 24 m) with maximum limit of 200 ft (61 m) with gas, vapor, and fully compensated liquid or mercury filled systems. Case-compensated liquid and mercury systems are limited to 15 ft (4.5 m) and 50 ft (15 m), respectively
Temperature Ranges:	About a dozen glass-stem thermometer ranges are available between −40 to 110°F (−40 to 40°C) and 50 to 750°F (10 to 400°C). These ranges are similar to those of mercury-filled systems. Gas-filled systems are standardly available between −100 to 100°F and 400 to 1200°F (−100 to 100°C and 200 to 650°C). Vapor-filled systems have narrower ranges, from −20 to 120°F and 250 to 450°F (−30 to 50°C and 50 to 150°C)
Inaccuracy:	Glass stem: either 1% of full scale, or one scale division, which as a minimum is 1°F (0.5°C). Filled system accuracies are also limited to one scale division
Costs:	A multiangle glass-stem thermometer with stainless steel wetted parts, a 6 in. (150 mm) thermowell, and a 9 in. (225 mm) scale is $250. A 4.5 in. (113 mm) diameter direct-mounted filled thermometer with a 6 in. (150 mm) stainless steel thermowell is about $200, while a 12 in. (300 mm) diameter, flush-mounted unit with 125 ft (38 m) of armored stainless steel capillary can approach $1000
Partial List of Suppliers:	ABB Kent-Taylor; Ametek Inc. U.S. Gauge Div.; Anderson Instrument Co.; Brooklyn Thermometer Co.; Buffalo Gauge Inc.; Dresser Industries, Instrument Div.; Duro Instrument Corp.; Enercorp Instruments Ltd.; Fenwal Inc.; Fischer & Porter Co.; Foxboro Co.; Griffith Industrial; Honeywell Industrial Controls; ITT Barton Instruments Co.; Jumo Process Control Inc.; Marsh Instrument Co.; Marshalltown Instruments; Mercoid Div. of Dwyer Instruments; Moeller Instrument Co.; Moore Products Co.; Palmer Instruments Inc.; Princo Instruments Inc.; Tel-Tru Manufacturing; Thermal Instrument Co.; 3D Instruments; Trend Instruments Inc.; Trerice Co.; United Electric; Weiss Instruments; Weksler Instruments Corp.

In a glass-stem thermometer the volumetric change resulting from thermal expansion is interpreted into temperature. In a filled system, the pressure inside a closed thermal system is used to measure temperature. The reading of the pressure gauge is affected not only by the process temperature at the bulb but also by other factors that can introduce errors. These include: 1) ambient temperature variation, 2) barometric pressure variation, and 3) the hydrostatic head of the liquid in the capillary, if it is at a different elevation from the readout. In order to obtain accurate temperature readings, these error sources must be minimized, as will be explained in this section.

GLASS-STEM THERMOMETERS

The glass-stem thermometer was the first closed thermal expansion system and has been known since Gabriel Daniel Fahrenheit investigated the expansion of mercury in the eighteenth century.

Mercury or some other liquid (alcohol, pentane) fills the glass bulb (see Figure 6a) and extends into the capillary bore of the stem. The space above the mercury column to the sealed top is evacuated, but occasionally it may be filled with an inert dry gas, such as nitrogen, to increase the temperature range.

The expansion of pure mercury is .01%/°F (0.005%/°C) and very linear; therefore, the volume of the bulb must be about 10,000 times the volume of the capillary between two marks 1°F (0.56°C) apart.

Although bulb and capillary could be made from the same type of glass, it is more convenient to make the bulb from glass with a good stability factor and the capillary from glass which is easier to work. For accurate measurements, the capillary must be properly annealed after it is drawn to the correct bore. Uniformity of bore is desirable but not absolutely necessary if the thermometer is calibrated at a sufficient number of points.

The glass-stem thermometer can be made for a very narrow range in temperature. For instance, consider a clinical thermometer whose full range may be 96 to 102°F (35.6 to 38.9°C) and with a stem 4 in. (101.6 mm) long, active length. If the bulb volume is 0.03 cubic in. (.5 cc) the diameter of the bore is about .001 in. (.025 mm). The stem is frequently designed to magnify such a small column for easier readability.

Another design feature of some glass-stem thermometers, notably the clinical type, is a restriction purposely placed in the capillary which prevents the liquid from returning toward the bulb when the thermometer is removed from the warmer object. This creates a separation of the column in the stem. In this case, it is a desirable feature because it permits a ''highest point'' reading or peak picker. If the separation occurs inadvertently in any glass-stem thermometer, however, the result is an erroneous reading. The column can usually be rejoined by shaking or tapping.

The design of a glass-stem thermometer requires that the filling material be a liquid over the entire range of temperature desired. Mercury is most suitable and can be used from its freezing point (−38°F, or −39°C) up to nearly its boiling point (over 1000°F, or 538°C). At this upper limit the space above the mercury column must be under great pressure with an inert gas to prevent evaporation of mercury from the top of the column. Alcohol and a few other hydrocarbons may be used for low temperatures. Colorfast dyes are usually added to these liquids to increase visibility.

To minimize accidental breakage a metallic thermowell is sometimes used to protect the bulb. This has no effect on the accuracy but may reduce the speed of response. Figure 6b illustrates the industrial packaging of the mercury or liquid-in-glass thermometer. These straight scale units have been ruggedized for mechanical strength and are still used in HVAC (heating, ventilation, and air conditioning) applications. In most other industries their use has been discontinued, mostly because of the concern about the toxicity of mercury.

Advantages of liquid-in-glass thermometers are low cost, simplicity, and long life if treated properly. Disadvantages are difficult reading, confinement to local measurement, and nonadaptibility to recording or automatic control. They also break very easily.

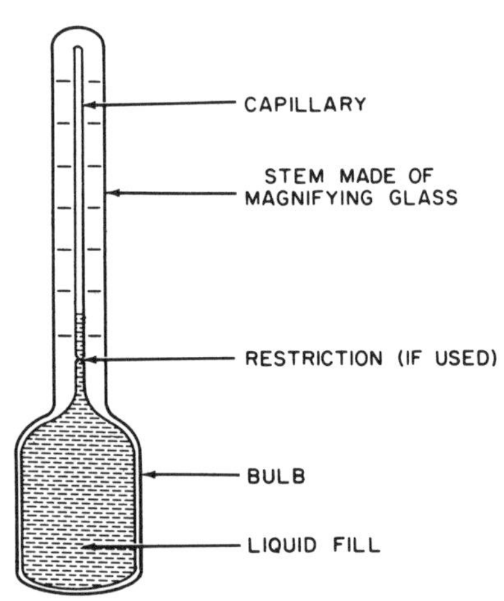

FIG. 6a
Liquid-in-glass thermometer.

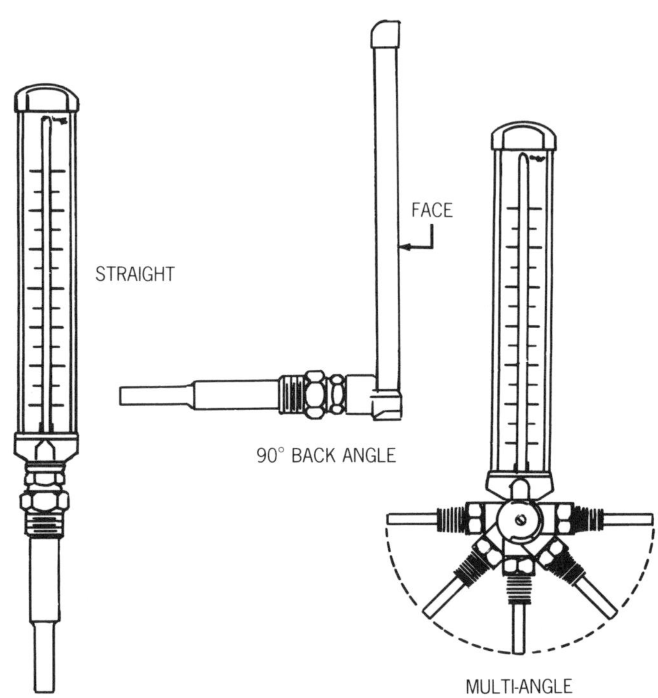

FIG. 6b
Industrial glass-stem thermometers. (Courtesy of Taylor Thermometer Corporation of America)

TABLE 6c

Thermal Systems Compared

SAMA Classification	Type Compensation	Fill	Measured Temperature °F (°C) Max.	Measured Temperature °F (°C) Min.	Span °F (°C) Max.	Span °F (°C) Min.	Scale Linearity	Speed of Response (No Well)	Elevation Head Effect	Overrange Capabilities	Superior Features	Less Desirable Features
I	None	Liquid	600 (315)	−125 (−87)	300 (167)	30 (17)	Linear Except at Low Temps.	6–7 Sec.	Significant; Must Be Compensated	About %100	Smaller Bulb Sizes; Narrower Spans; Lower Cost	Shortest Capillary; Compensation More Difficult and Required
IA	Full											
IB	Case											
IIA	Not Required	Vapor	Only Above Ambient to 600 (316)		400 (222)	60 (33)	Scale Divisions Increase with Temp. Increase	4–5 Sec. Except When Passing Through Ambient	Appreciable Compensation	Small or None	Low Cost; Long Capillary Lengths Available; Ambient Temperature Comp. Not Req'd; Fast Speed of Response	Problem with Cross-Ambient; Non-linear Scale No Overrange Capacity; Elevation Error Possible
IIB			Only Below Ambient to −40 (−40)						None			
IIC			Above and Below Ambient Not Through						Not Usable; Can't Compensate			
IID			Above, Through and Below Ambient						Same as IIA			
IIIA	Full	Gas	1400 (760)	−450 (−268)	1200 (667)	120 (67)	Linear Except Very Low Temps.	4–7 Sec.	None	150–300% Varies with Span	No Head Effect; Greatest Overrange; Greatest Range of Temperatures; Good for Averaging	Large Bulb Size; Widest Spans Req'd; Least Power for Control
IIIB	Case											
VA	Full	Mercury	1200 (649)	−40 (−40)	1200 (667)	50 (28)	Linear	4–5 Sec.	Not Significant Up to 30 Ft. (9.1 m)	100% or More	Most Linear Scale; Least Difficult to Compensate; Most Power for Control Elements; Small Bulb	Objection to Mercury on Accidental Breakage; Compensation Required
VB	Case											

FILLED THERMAL SYSTEMS

A filled thermal system is basically a pressure gauge connected by small bore tubing to a bulb acting as the temperature sensor. The whole system is gas-tight, and filled with an appropriate confined gas or liquid under pressure.

There are many different types of filled systems, each having certain peculiarities which give it advantages over others.

The Scientific Apparatus Makers Association (SAMA) has classified filled-system thermometers into four major classes according to filling material. Table 6c has been prepared to compare some of the properties of the various thermal systems.

The basic components of a filled-bulb capillary thermometer are illustrated in Figure 6d. As mentioned previously, the use of mercury filling has been discontinued in most industrial applications because of health concerns. The use of both mercury and liquid fillings have also lost ground because of the expense associated with compensating for the ambient effects on the capillary and because of the errors caused by elevational differences between bulb and readout. Still extensively used are the gas and vapor fillings, but these too are limited in their usefulness. The gas-filled bulb is large, as is the temperature span required to operate it. The vapor-filled system is also limited due to its nonlinearity and its potential for problems caused by cross-ambient operation or by errors due to elevation.

BULBS, WELLS, AND CAPILLARIES

The temperature-sensitive element, the bulb, comes in many sizes and shapes to handle the many different applications. It is good practice to use the largest bulb which will do the job. This will cut down on ambient temperature errors, and permit smaller spans and longer capillaries.

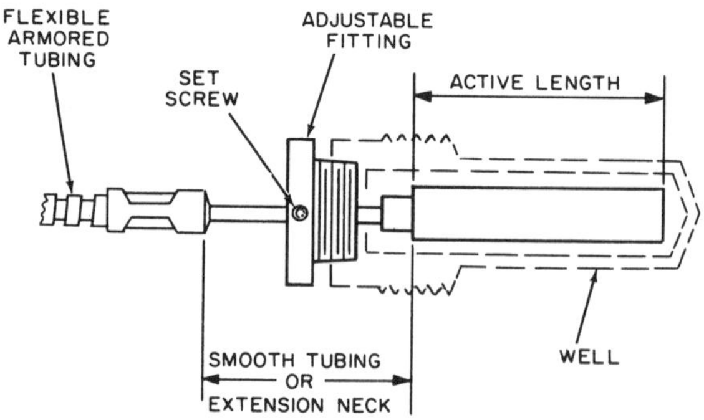

FIG. 6e
Plain bulb with adjustable fitting and well.

Plain bulbs are used where the measured medium is not under pressure and will not harm the bulb material. If this is not the case then a separate well to protect the bulb from the process medium is needed. This, of course, will slow down the response time even more than the 4 to 7 seconds speed of response range given in Table 6c.

The speed of response generally doubles with the doubling of bulb diameter and tends to be the fastest with vapor or gas and the slowest with liquid filling, with mercury and Class IID vapor filling in the middle. When thermowells are added to the bulb, speed of response slows as follows: 12 to 24 seconds for 0.25 in. (6 mm), 20 to 35 seconds for ⅜ in. (9 mm), 25 to 50 seconds for 0.5 in. (13 mm), and 40 to 75 seconds for ¾ in. (19 mm) diameters. Figure 6e illustrates a filled bulb installed in a thermowell.

Higher speed of response can be obtained with a long, thin, bendable bulb used to sense the average temperature in large areas. Another way of presenting a long bulb is by coiling it. For use in low-velocity gas flow temperature measurement the coil is set at the factory and cannot be uncoiled (Figure 6f).

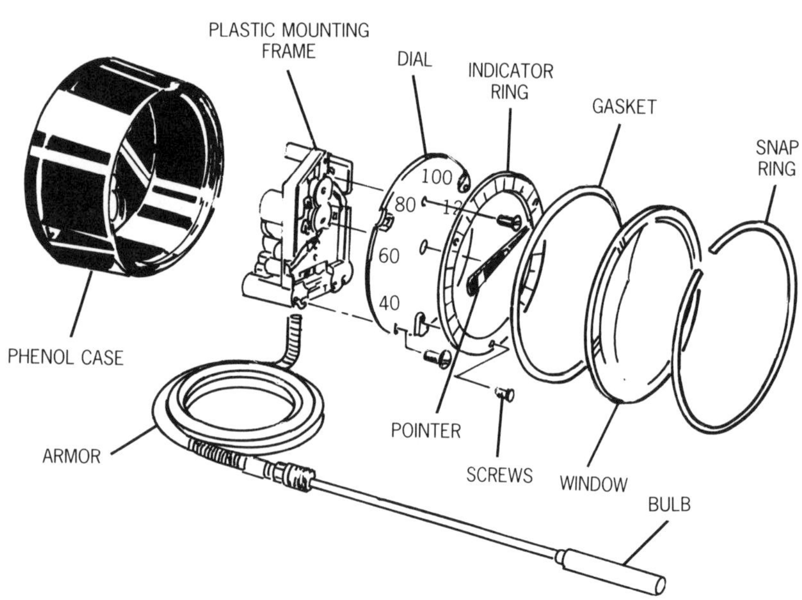

FIG. 6d
The components of a filled-bulb capillary thermometer. (Courtesy of Ashcroft Div. of Dresser Industries Inc.)

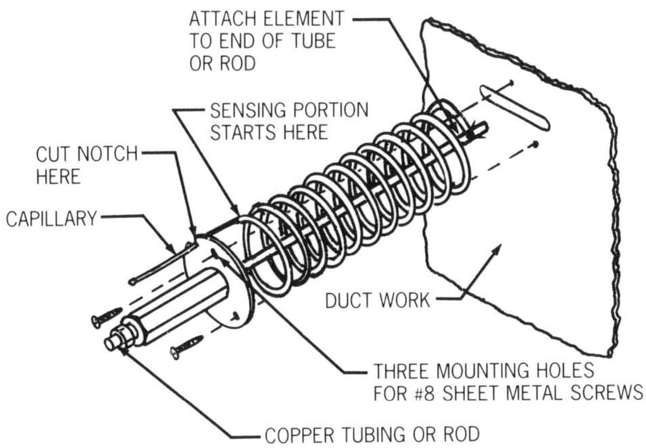

FIG. 6f
Averaging capillary for air duct installation. (Courtesy of Johnson Controls)

Most bulbs are made of stainless steel, which is relatively inert and will withstand high temperatures. Other materials are readily available.

The relatively fragile thin-wall capillary should be protected by a flexible armored stainless steel or PVC-covered bronze tubing. An extension neck to the bulb prevents the tubing from being immersed directly in the measured medium. Bendable smooth steel tubing might also be used for capillary protection.

Class I, Liquid-Filled Systems

These systems are completely filled with a liquid (other than mercury) and operate on the principle of liquid expansion with increase of temperature (Figure 6g). The filling fluid is usually an inert hydrocarbon, such as xylene (C_8H_{10}), which has a coefficient of expansion six times that of mercury and makes smaller bulbs possible. Other liquids (even water) are

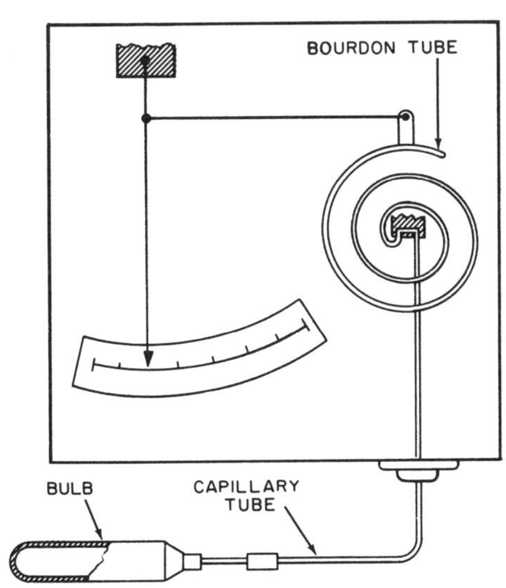

FIG. 6g
Liquid-filled capillary-type thermometer.

sometimes used. The criterion is that the pressure inside the system must be greater than the vapor pressure of the liquid to prevent bubbles of vapor from forming in the spiral. Also, the liquid should not be allowed to solidify even in storage; otherwise, the calibration may be affected.

The minimum operating temperature is usually set by the freezing point of the filling liquid, which is usually between −100 and −350°F (−75 and −210°C). The maximum operating temperature is set by the point at which the filling liquid is no longer stable, usually around 600°F (315°C). The minimum span range is a function of bulb size; the maximum, of linearity. With larger bulbs the span can be narrowed to 22 to 45°F (12 to 25°C), while the maximum limit on the span is about 300°F (167°C) due to the nonlinearities caused by expansivity and compressivity of the filling liquids.

The maximum temperature which the bulb can be exposed to without damage is defined as the allowable overrange of the system. Overrange is usually expressed as a percentage of the span over the full range value; in the case of liquid filling, overrange is around 100%. Therefore, if the range of a liquid-filled thermometer is 100 to 300°F and its overrange protection is 100%, its bulb can be exposed to a maximum temperature of 500°F.

Class II, Vapor Systems

The pressure element, capillary, and bulb of a Class II system have the filling medium in both the liquid and vapor form. The interface between the two must occur in the bulb, and this will move slightly with temperature, affecting the pressure.

The pressure within the system is a function of the vapor pressure of the filling fluid at the operating (bulb) temperature. The filling fluid is usually so selected as to give a 100 PSIG (6.9 bars) change in its vapor pressure as the bulb temperature goes through the span of the thermometer. This much change is necessary to keep the error due to barometric pressure changes to less than 0.5% of full scale. The filling fluids used include methyl chloride, sulfur dioxide, butane, propane, hexane, methyl ether, ethyl chloride, ethyl ether, ethyl alcohol, and chlorobenzene. Each has a different vapor-pressure–temperature relationship. For lower temperatures ethane filling can be used; it changes its vapor pressure from 20 to 600 PSIG (1.4 to 41 bars) as the temperature rises from −100 to 85°F (−73 to 29°C). For higher temperatures one might use ethyl chloride, which goes through the same vapor pressure change as its temperature rises from 100 to 350°F (38 to 177°C).

Overall, the minimum temperature at which vapor fillings can be used is around −40°F (−40°C), while the maximum is about 600°F (315°C). The maximum temperature is limited by the critical point of the fill, while the minimum limit is a consequence of the loss of reading sensitivity, as the vapor pressure changes less per unit temperature change at low temperatures. The nonlinear nature of the vapor-filled thermal systems is generally considered to be a disadvantage, although if a better reading sensitivity toward the top of the scale is required, it can be a desirable feature.

The speed of response of vapor-filled systems is generally in the range of 1 to 10 seconds. It is faster than the liquid or mercury fills and as fast as the gas-filled systems in most of its configurations, except the Class IID design, which is a little slower.

The overrange limit of the vapor-filled systems is small, because the vapor pressure tends to rise exponentially with temperature. Under some very unique sets of conditions it is possible to obtain higher overrange by designing the thermal system so that all liquid will be exhausted at a temperature slightly over the maximum required reading.

Class IIA System In a thermal system, which is filled with a liquid that is in equilibrium with its own vapors, the liquid will always situate itself at the cold end and the vapor at the hot end of the system. As shown in Figure 6h, in the Class IIA thermal system the bulb is mostly filled with vapors, while the capillary and the readout instrument are liquid-filled. This type of system can be used only if the temperature at the bulb is *always* above the ambient temperature around the readout instrument. As the process temperature rises, more liquid is vaporized, which in turn increases the vapor pressure in the bulb and in the bourdon tube of the readout instrument. In this system, the changes in ambient temperature around the capillary will change the volume of the filling fluid and will necessitate using the bulb as a thermal expansion chamber. Consequently, the longer the capillary tubing, the larger the required bulb volume becomes. This tends to limit the maximum capillary length to about 75 ft (23 m).

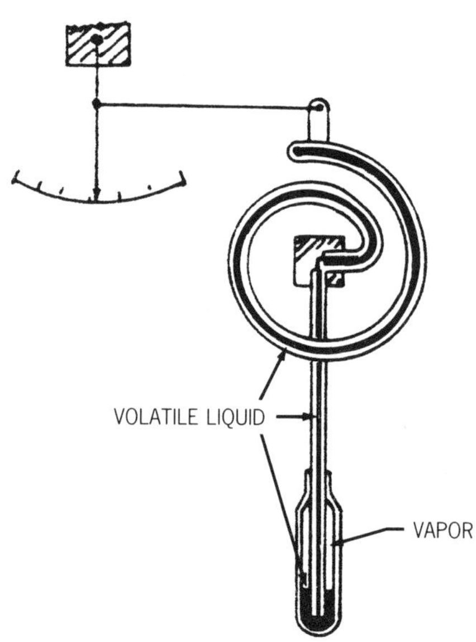

FIG. 6h
Class IIA vapor filling.[1]

Class IIB System As shown in Figure 6i, in a Class IIB thermal system, all the filling liquid is in the bulb, while the rest of the system is vapor-filled. This type of configuration can be used only if the temperature at the bulb is always *lower* than the ambient temperature around the capillary or at the readout device. If this was not the case, the liquid would start boiling in the bulb and recondensing in the bourdon tube. This transfer of liquid cannot be completed because there is not enough liquid to fill the bourdon tube and the capillary; while the transfer is taking place, the temperature readings are meaningless. The Class IIB system requires the smallest of the vapor-filled bulbs as it does not have to act as a thermal expansion chamber.

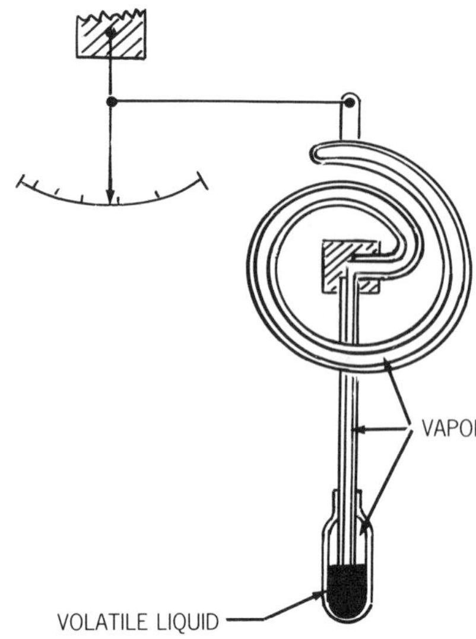

FIG. 6i
Class IIB-type vapor filling.[1]

Class IIC System Figure 6j illustrates the Class IIC thermal system, which is capable of operating as a Class IIA (right)

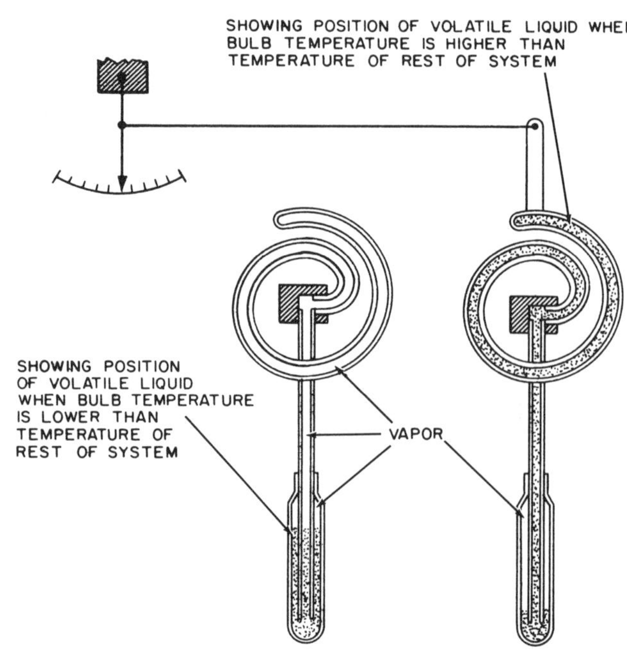

FIG. 6j
Class IIC system.

or a Class IIB (left), because its bulb is large enough to accept all the liquid filling from the bourdon tube and capillary. This obviously means that the bulb required for this design is larger than for either of the two previously described systems. This design can be used when the process temperature is either above or below the ambient, but *it cannot be used when it might cross the ambient temperature,* as it takes some time for the filling liquid to transfer itself to the cold end of the thermal system and to settle down after the liquid/vapor interface has been reversed.

Class IID System In the Class IID system there are two filling liquids: one is volatile and is always situated in the bulb, while the other is nonvolatile and fills the bourdon tube, capillary, and part of the bulb (Figure 6k). The purpose of the nonvolatile liquid is only to transmit the vapor pressure at the volatile liquid/vapor interface, which is always confined within the bulb. In this design the process temperature can be anything, including the crossing of the ambient. This configuration requires an even larger bulb than that needed by the Class IIA design.

Class III, Gas-Filled Systems

The operating principle for gas-filled systems is that in a perfect gas confined to a constant volume the pressure is proportional to the absolute temperature. The gas is not perfect and not all at the same temperature nor is the volume constant, but variances are small enough so that a measurement of pressure can be used to indicate temperature.

Nitrogen is the favorite fill for a Class III system because it is inert and inexpensive. It does react somewhat with the steel bulb material at temperatures exceeding 800°F (427°C), and it does act less like a perfect gas at extremely low temperatures. Under these conditions helium should be used. Different ranges are obtained by selecting the correct filling pressure.

In general, bulbs should be as large as practical to lessen the influence of temperature variations along the capillary. One way of avoiding long capillaries is to terminate a short capillary at a small diaphragm chamber. The force due to gas pressure on the diaphragm causes it to compress the spring. This motion is amplified and used to regulate another pressure which is transmitted to the spiral. This arrangement, though more expensive, permits much smaller bulbs than could otherwise be used.

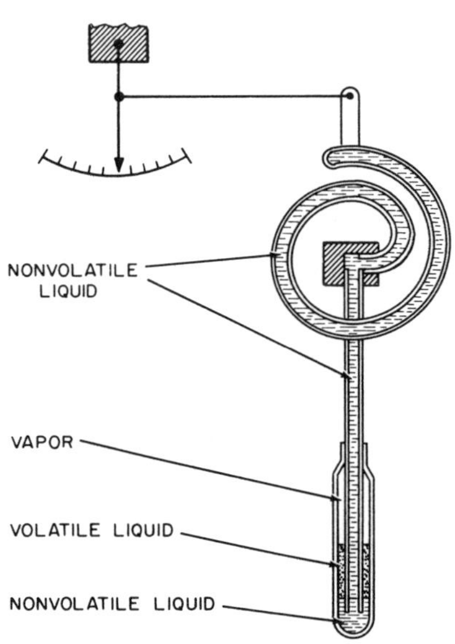

FIG. 6k
Class IID system.

Gas-filled systems approximate Charles's law (absolute pressure of a confined gas is proportional to its absolute temperature) by keeping the bulb volume relatively large compared to the rest of the system. Class III systems are primarily used for low and high temperatures. On the low side they are limited by the critical temperature of the filling gas (usually nitrogen or helium) corresponding to −450°F (−268°C), and on the high side by the temperature limits on the bulb materials, usually corresponding to 1400°F (760°C). The maximum span can be 1200°F (667°C) and is limited only by nonlinearities due to mass flow from the bulb. The minimum span is limited by the pressure at which the bourdon tube becomes overstressed (usually designed for 400 PSIG, or 28 bars). Therefore, the minimum span with conventional systems is about 400°F (222°C), while with helically wound thin-tube bulbs it can be reduced to 120°F (67°C) or even less.

The speed of response of gas-filled systems is usually good. The time constant is only 1 and 4 seconds because the ratio between bulb mass and surface area tends to be favorable. Class III filling usually can provide 150 to 300% overrange protection as the maximum temperature is limited only by the permissible pressure and temperature ratings of the bulb. The gas-filled bulb is ideal for measuring the average temperature over large areas, such as in dryers or ovens. This is the case because the bulb volume is not critical and capillary-type bulbs (Figure 6f) can cover large areas.

Class V, Mercury-Filled Systems

Mercury is a liquid and in this respect Class V systems are similar to Class I. The two are separated because of the unique characteristics of mercury and its importance as a temperature-measuring medium. Mercury provides rapid response, accuracy, and plenty of power for operating control elements. Pressures within the working system are relatively high—as much as 1200 PSIG (83 bars) for the higher temperatures, dropping to 400 PSIG (28 bars) at the low-temperature end of the range. This high pressure cuts down on any head effect error (difference in elevation between bulb and measuring instrument).

Mercury-filled systems can detect temperatures between the freezing and the boiling points of mercury, or from −40 to 1200°F (−40 to 649°C). Mercury fillings also provide the widest range of spans, from 50 to 1200°F (28 to 667°C). The speed of response of mercury-filled systems is faster than that of the liquid-filled ones but slower than that of gas or vapor systems. The Class V systems can have an overrange protection of at least 100%.

AMBIENT-TEMPERATURE COMPENSATION

The measured variable in a filled thermometer is the total internal pressure. This pressure is the result of two factors—the temperature around the bulb and the ambient temperature around the rest of the system. The purpose of the installation is to measure the process temperature around the bulb, and therefore it is desirable to eliminate the effects of ambient-temperature variations on the total reading. The error introduced by ambient-temperature variations is different for the different types of fills, and it also increases as the bulb or span gets smaller or as the capillary length is increased.

When the ambient temperature effects are compensated for in both the capillary and in the readout instrument, the design is called fully compensated. Full compensation is designated by the letter "A" being added to the Class designation number I, III, or V. When the case and the capillary is at the same ambient temperature and the length of the capillary is relatively short, it might be sufficient to leave the capillary portion uncompensated and provide case compensation only. When a Class I, Class III, or Class V system is case-compensated, the letter "B" is added to the designation.

Class II (vapor-filled) systems do not need any compensation because the liquid/vapor interface is always within the bulb. In the case of Class IIA fillings, it is important to keep all sections of the capillary at a temperature which is *below* that at the bulb; otherwise, the fill fluid could vaporize in the capillary and introduce substantial errors.

Figure 6l describes a method of fully compensating Class I, Class III, or Class V systems. Here the spiral pressure sensor and the capillary are both duplicated. The compensating capillary is closed, and its length is the same as the active capillary without the bulb. The two capillaries are run parallel to each other and the two bourdon spirals are connected in such a way as to cancel the ambient effects.

In the case of mercury-filled Class V systems, full compensation can be provided without the expense of duplicating most of the thermal system. This cost-saving approach is illustrated in Figure 6m where an Invar wire is inserted inside the single active capillary. Invar is a nickel alloy which has a low coefficient of thermal expansion. Therefore it is possible to select the diameters of the Invar wire and of the capillary in such a manner that an increase in ambient temperature will increase the annular space between the wire and the capillary by just about the same amount as the volumetric expansion of mercury. Therefore, a change in ambient temperature around the capillary does not change the pressure inside the thermal system.

When both the capillary and the case are at the same ambient temperature and the capillary volume is very small relative to the total volume of the system, it is possible to leave the capillary uncompensated and limit the compensation to the case. Case compensation (Class IB, Class IIIB, or Class VB) can be achieved by the addition of a bimetallic spring which generates a force that is nearly equal to the one caused by the change in ambient temperature (Figure 6n). Because the volume of the bourdon spiral varies as the internal pressure changes with the measured temperature,

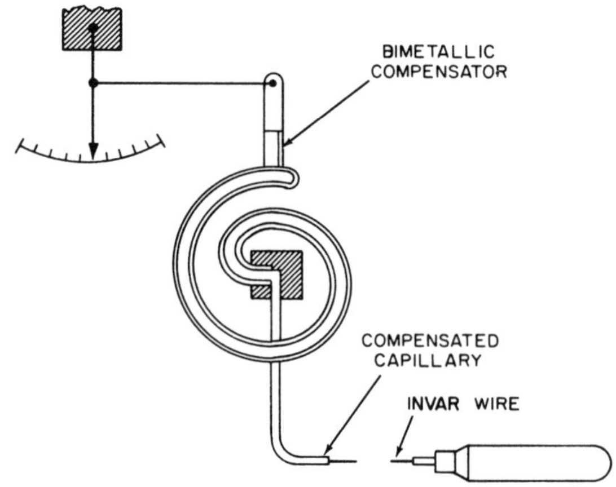

FIG. 6m
Class V full compensation with internal wire.

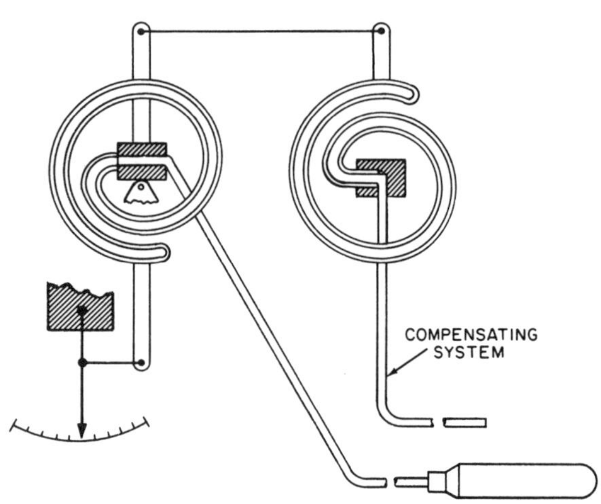

FIG. 6l
Full compensation.

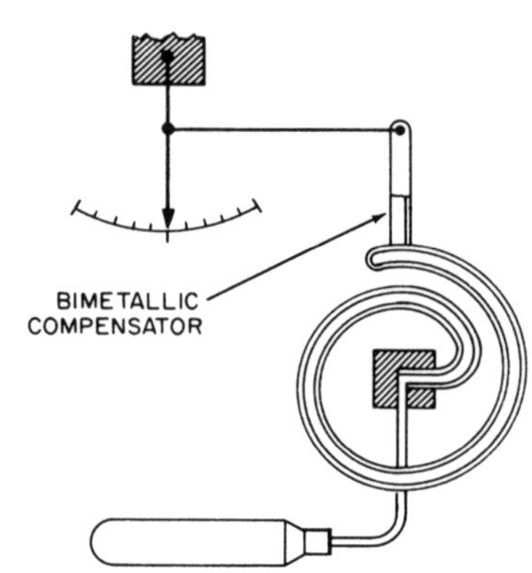

FIG. 6n
Case compensation.

the case compensation provided by the bimetallic spring is effective at only one bulb temperature. The bimetallic spring is usually set at midrange, resulting in undercompensation at the top and overcompensation in the bottom half of the range.

Uncompensated operation (Figure 6g) is possible with liquid-filled Class I systems when the capillary is very short and with all types of Class II vapor-filled systems. Class III and Class V units must always be compensated. In gas-filled Class III systems, the capillary error rises rapidly with bulb temperature and can be reduced by increasing bulb volume and/or span.

EFFECTS OF BULB ELEVATION

In a filled system a pressure gauge is used to measure temperature. If the thermal bulb is above the readout instrument, the hydrostatic head of the filling fluid will be added to the total pressure at the spiral bourdon element. If the bulb is below the case, the hydrostatic head will be subtracted from the total pressure. Therefore, when the bulb is high the thermometer will have a positive error, and when it is low, it will have a negative error. The relative size of the error is a function of the size of the hydrostatic head relative to the total pressure.

In the case of liquid- or mercury-filled systems (Class I and Class V), the hydrostatic head caused by the elevational difference between the case and the bulb is a constant and therefore can be zeroed out. Depending on whether the bulb is above or below the case, the hydrostatic head must be subtracted or added at the readout bourdon tube. This is done with suppression or elevation springs (respectively). The amount of elevation or suppression that is available is not unlimited. Due to the high density of mercury in the Class V systems, the hydrostatic head due to bulb elevation is much greater than with other fillings. Therefore, with mercury-filled systems the maximum elevational difference that can be tolerated is about 30 ft (9 m). Needless to say, if the readout of the bulb is moved, the system must be rezeroed.

In the case of Class I and Class V systems, if the bulb is above the case, the manufacturer must also pressurize the filling liquid by something more than the hydrostatic head, in order to make sure that the pressure in the bulb does not drop to zero. This increase in the initial filling pressure reduces the overrange capabilities of the system.

For Class IIA, Class IIB, and Class IID vapor-filled systems, correction of the elevational error is similar to correction for Class I and Class V, using elevation or suppression to compensate for the hydrostatic heads and thereby to rezero the system. In the case of Class IIC systems the height of the liquid column in the capillary varies with temperature and therefore cannot be zeroed out as a fixed quantity. Consequently, Class IIC systems *should not be used* when the case and bulb are at different elevations.

In the case of Class III gas filling, the density of the gas (and therefore its hydrostatic head) is too small to cause any errors. Therefore no correction is needed to protect against bulb elevational errors.

BAROMETRIC ERRORS

The bourdon spiral is a differential pressure sensor. Its deflection is a function of the difference between the pressure of the filling fluid on the inside and the atmospheric pressure on the outside. Because the atmospheric pressure is not a constant, its changes can also result in readout errors. The barometric pressure can change by about 0.5 PSIG (25 mmHg). This may or may not be large enough to introduce an error, depending on the filling pressure. The pressure in the readout spiral varies with the type of filling fluid. In Class I and Class V systems it is high enough to make the barometric effect insignificant.

Class II vapor-filled systems are generally designed for a minimum of 100 PSIG (6.9 bars) pressure change across their span, and therefore the barometric effect can result in a 0.5% error. Class III gas-filled systems are usually designed to produce a 400 PSIG (28 bars) pressure change across their ranges. Therefore the barometric effect can result in an error of about 0.1%.

CONCLUSIONS

The filled systems are superior to bimetallic elements but are generally inferior to electronic thermometers. Their advantages include rugged, simple, self-contained, inexpensive, easily maintained construction and acceptable sensitivity and accuracy without need for auxiliary power sources. They are inherently explosion-proof and can operate displays or recorders at some distance from the measurement using only hand-wound spring drives.

Their disadvantages include the bulky bulbs, which are space-consuming and result in a relatively slow speed of response (about 20+ seconds with well), inability to provide narrow spans, limitation on the distance between bulb and readout (usually under 200 ft, or 60 m), and relatively difficult maintenance of the pressurized thermal elements, which is usually done at the factory. Filled systems are not suited for high-temperature measurement and can be easily misapplied if the user is not familiar with the unique characteristics of each system.

References

1. Nelson, R.C., "Temperature Measurement by Filled System," *Instruments and Control Systems,* May 1961.

Bibliography

Carlson, D.R., "Transmitting a Remote Temperature," April 1989.

Considine, D.M., *Encyclopedia of Instrumentation and Control,* New York: McGraw-Hill, 1971.

''Filled-Systems Thermometers,'' *Measurements and Control,* September 1992.

Hashemian, H.M., et al., ''Assuring Accurate Temperature Measurement,'' *InTech,* October 1989.

Hormuth, G.A., ''Ways to Measure Temperature,'' *Control Engineering,* Reprint No. 948, 1971.

Peterson, W., ''Choosing the Right Temperature Transmitter,'' *InTech,* April 1991.

Plumb, H.H., ''Temperature, Its Measure and Control in Science and Industry,'' 5th Symposium on Temperature, held in 1971 by NBS, AIP, and ISA.

When, S., ''Liquid-in-Glass Thermometers, Still Well and Indicating,'' 1991 ISA Conference, Paper #91-0308.

7 Integrated Circuitry (IC) Transistors and Diodes

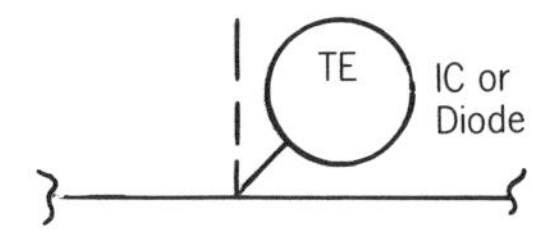

Flow Sheet Symbol

B. G. LIPTÁK (1993)

Applications:	IC sensors are most often used in near-ambient temperature measurement; germanium diodes are often used in extreme cryogenic applications, and silicon diodes are used from −60 to 300°F (−51 to 149°C)
Ranges:	IC sensors, from −65 to 300°F (−54 to 149°C); silicon diodes, from −60 to 300°F (−51 to 149°C); germanium diodes, from −420 to 110°F (−251 to 43°C). Some diodes can cover the range of 1.4 to 475°F (−271.8 to 202°C)
Outputs:	Some IC sensors generate a linear output with a slope of one microampere per degree C, others produce an output of 10 millivolts per °C
Linearity:	1% of range
Inaccuracy:	For IC sensors the error is from 1 to 9°F (0.5 to 5°C), with sensitivities on narrow spans approaching 0.1°F (0.06°C). For calibrated diodes the error is 0.2°F (0.1°C) for temperatures above ambient and 0.2% of range below that. Some calibrated diodes are accurate within 0.05°C (0.1°F) between 2 to 305°K (−271 to 32°C)
Costs:	The basic IC transistor elements can cost less than $10. Uncalibrated diode sensor costs range from $75 to $400; calibration can cost an additional $2-300. A calibrated diode with packaging and mounting for cryogenic service costs $520. Indicating readouts for diode elements cost about $800.
Partial List of Suppliers:	Analog Devices Inc.; CI Systems Inc.; Doric Scientific; Dynalco Controls/Mark Controls; Elan Technical Corp.; Elcontrol USA Inc.; Elmwood Sensors Inc.; Lake Shore Cryotronics Inc. (diodes); Motorola Semiconductor Products; National Semiconductor; Omega Engineering Inc.; Onicon Inc.; OPTO 22; Raychem Corp. (multipoint IC tape); S-Products Inc.; Syscon International Inc.; Weschler Instruments; Wilkerson Instrument Co.; Yokogawa Corp. of America

Diodes and integrated circuitry (IC) transistors have been available as temperature sensors for several decades, but their use in routine industrial applications is relatively recent. The silicon (and germanium) transistors are small and inexpensive, but before they can be used as part of an industrial system (such as intelligent thermostats), they need to be packaged. One of the first applications of the IC transistor was to provide cold junction compensation for thermocouple circuits. They are also used in multipoint temperature-sensing cables and in other applications. Today both the diode and the IC sensors are available as fully packaged, off-the-shelf units; some of them are also integrated with microprocessors, resulting in a ''smart'' sensor. Diodes and transistors can both provide sensitivities approaching 0.1°F (0.06°C) on narrow span applications, but in the case of the diodes calibration is needed to achieve that sensitivity.

INTEGRATED CIRCUIT TEMPERATURE SENSORS

Transistors are sensitive to temperature variations. It has been found that if two identical transistors are operated at a constant ratio of collector current densities, then the difference in their base-emitter voltages will be directly proportional to absolute temperature. Therefore, as temperature decreases the base bias must be increased to maintain the collector current constant. The base bias voltage is usually converted to a current by a low-temperature coefficient thin film resistor. The temperature-proportional characteristic is what allows IC transistors to produce output signals that are proportional to absolute temperature (Figure 7a). The IC temperature sensors are available in both voltage and current output configurations. The current output units are usually set for a one microampere output change per degree Kelvin (Celsius), while the voltage output configuration generates 10 millivolts per degree Kelvin. If a digital voltmeter is used as the readout, the current output is detected as the voltage drop through a 10,000 ohm resistor (Figure 7b). The temperature-proportional characteristics of IC sensors are highly linear and in the range of −67 to 302°F (−55 to 150°C) it is superior to all electronic sensors, including RTDs.

Because they produce an analog voltage that is proportional to temperature, IC transistors have been used for some time for cold junction compensation in thermocouple cir-

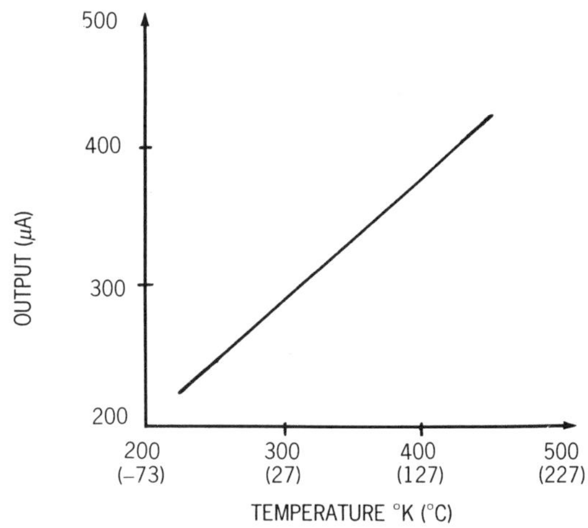

FIG. 7a
The temperature-proportional characteristics of the silicone transistor. (Courtesy of Analog Devices Inc.)

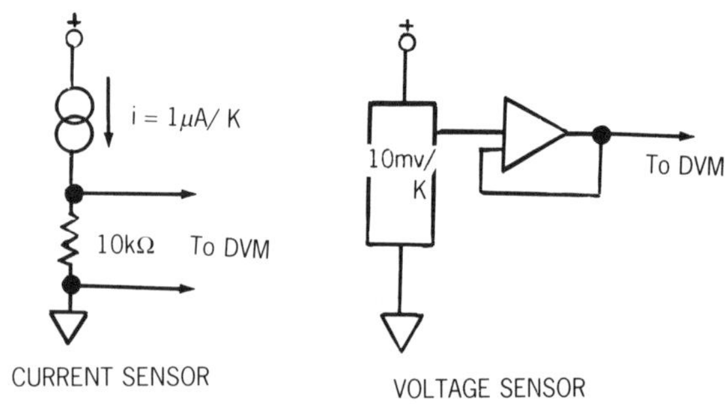

FIG. 7b
Current and voltage output type IC transistors. (Courtesy of Hewlett-Packard)

cuits. IC sensor modules have also been packaged in flat cables, where they are attached to the cable at regular (several feet) intervals. The cables can be as long as 10,000 ft (3050 m) and can have as many as 1000 sensing modules. These cables are economical ways of detecting pipe surface

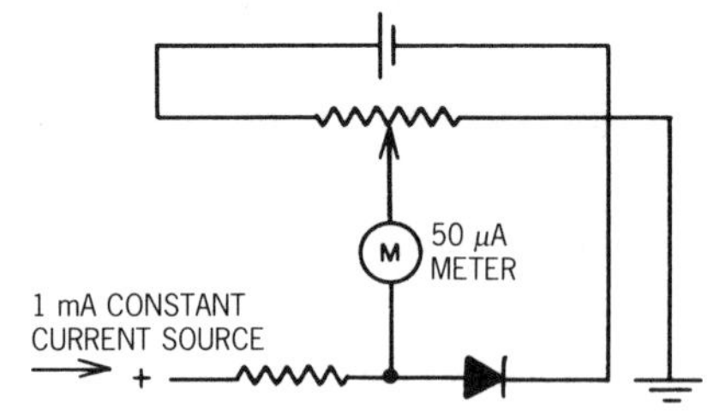

FIG. 7c
Simple silicon diode thermometer circuit.

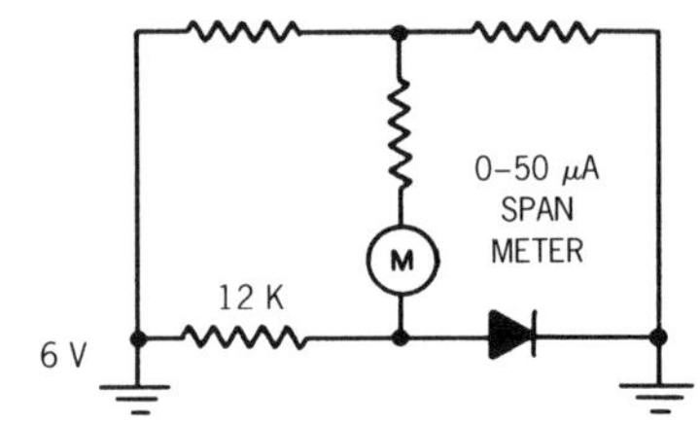

FIG. 7d
Germanium diode thermometer circuit.

temperatures or cold or hot spots. IC sensors are also being used in the HVAC (heating, ventilation, and air conditioning) industry and in other applications where their low cost and strong linear output outweighs their low accuracy and limited range. Other disadvantages of IC sensors include that they require an external power source (4 to 30 volts) and that they are fragile and subject to errors due to self heating.

DIODE-TYPE TEMPERATURE SENSORS

Diodes are highly sensitive and linear temperature sensors. Silicone and germanium diode temperature elements are available from −458 to 395°F (−272 to 202°C). They are accurate to 0.2°F (0.1°C) for temperatures that are above ambient and to 0.2% of their full scale range below that. Figure 7c depicts the diode-based thermometer circuits of a silicone diode with a 1 milliampere constant current source,

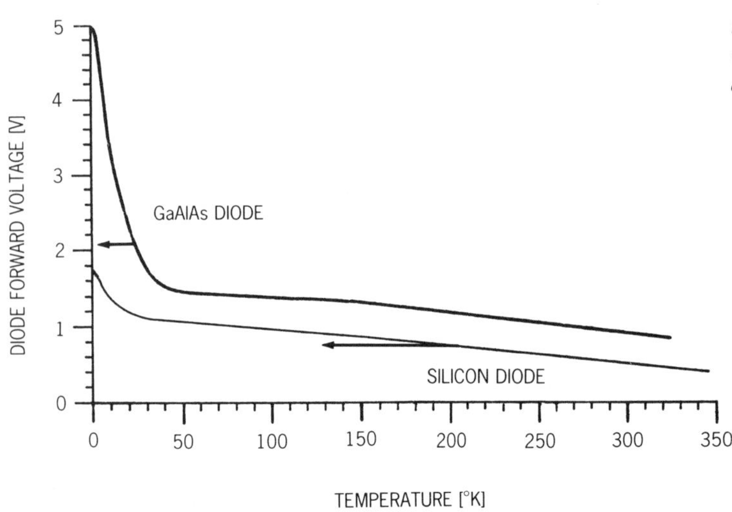

FIG. 7e
Characteristics of silicone and gallium-aluminum-arsenide diodes.[1]

and Figure 7d shows the circuits of a germanium diode with a 6 volt power supply. The current through the silicon diode thermometer readout could be anywhere from 10 microamperes to 10 milliamperes, but because higher currents result in higher errors due to self-heating, the currents are usually kept between 100 and 500 microamperes. The meters used with germanium diodes usually have a range of 0 to 50 microamperes. Because of the low source impedances, a simple microampere indicating meter can be used. This sensor can also be used to detect small temperature differences. The main advantages of the diode-type sensors are their high accuracy, particularly at cryogenic temperatures; their small size; their low cost; and their good linearity (Figure 7e). The diodes are small enough for most applications, but where sensor size is a problem, microdiodes can be used. One of the disadvantages is that the variations between diodes require a calibration procedure for absolute accuracy, which also increases their costs.

References

1. Courts, S.S., Holmes, D.S., et al., ''Cryogenic Thermometry—An Overview,'' from *Applications of Cryogenic Technology,* Vol. 10, J.P. Kelley, ed., New York: Plenum Press, 1991.

Bibliography

Courts, S.S., et al., ''Cryogenic Thermometry, An Overview,'' American Physical Society Conference, 1991.

''Integrated Circuit and Diode Thermometry,'' *Measurements and Control,* February 1991.

Kelley, J.P., ed., *Applications of Cryogenic Technology,* Vol. 10, New York: Plenum Press, 1991.

Krause, J.K., and Dodrill, B.C., ''Measurement System Induced Errors in Diode Thermometry,'' *Review of Scientific Instruments* 57 (4), 1986, pp. 661–665.

Krause, J.K., and Swinehart, P.R., ''Demystifying Cryogenic Temperature Sensors,'' *Photonics Spectra,* August 1985, pp. 61–68.

Langan, P.E., ''Calibration Ensures Accurate Temperature Measurement,'' *InTech,* October 1991.

Peterson, W., ''Choosing the Right Temperature Transmitter,'' *InTech,* April 1991.

''Practical Temperature Measurements,'' Hewlett-Packard, Application Note 290.

Preston-Thomas, H., ''The International Temperature Scale of 1990 (ITS-90),'' Metrologia, 27, 3–10, 1990.

Sparks, L.L., ''Temperature, Strain, and Magnetic Field Measurements,'' in *Materials at Low Temperatures,* R.P. Reed and A.F. Clark, eds., American Society of Metals, Metals Park, pp. 515–571.

White, G.K., *Experimental Techniques in Low Temperature Physics,* Oxford: Clarendon Press, 1979.

8 Miscellaneous Temperature Sensors

T. J. CLAGGETT, R. W. WORRALL (1969, 1982) B. G. LIPTÁK (1993)

In this section several unrelated methods of temperature measurement will be briefly discussed. Some of these are still in the experimental stage, while others have not been exploited commercially because they are not yet competitive with the more common sensors. This fast-changing field is so broad that it is not easy to provide complete coverage.

SELF-MEASURING DEVICES

Sometimes temperatures can be measured by allowing a material to serve as its own thermometer. This is done when determining the pressure of a confined gas to indicate its mean temperature, or the resistance of heater coils for the same purpose.

Actually, any property having a consistent rate of temperature variation will serve to indicate temperature. The frequency of the chirp of crickets on a summer night is an indication of their temperature environment. The rate at which a viscous substance such as oil drips through a small hole in the bottom of its container is just as much an indication of its temperature (provided the time-temperature relationship is known) as would be obtained by inserting a glass-stem thermometer.

In addition to such self-measuring devices, there are also many new configurations of older systems. For example, the bimetallic spring can now be used as an optical temperature switch. In this design the bimetallic element is upwardly convex until the set temperature is reached. At that point it becomes downwardly convex, which blocks an optical light path. The number of devices such as this which exploit old concepts in new ways is very large, and no attempt will be made to cover them here.

ACOUSTIC TIME DOMAIN REFLECTOMETRY

This thermometer operates on the principle that ultrasound pulses travel in solids at speeds which are a function of the temperature of the solids (Figure 8a). The measurement is made by detecting the time needed for the acoustic pulses to travel from the transducer to the impedance demarcation point (which may be the junction between the wire and the wall of the tank) and back to the transducer. This device is in the development stage and shows good potential, although some drift in the measurement has been reported. The rhenium sensor used in the system will probably require protection with some sheathing material in the more hostile environments.

CARBON RESISTORS

Commercially available carbon resistors have been used as temperature sensors in the cryogenic temperature area near

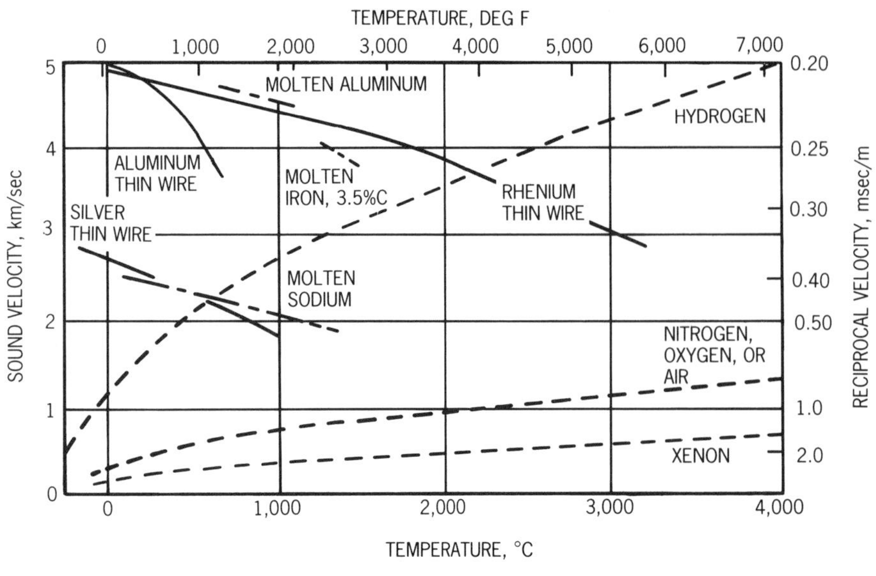

FIG. 8a
The velocity of sound is higher in thin wires when they are cold, and it drops as the wires are heated.[2]

absolute zero (from about −424°F, or −253°C, downward to below −458°F, or −272°C). The problems below −458°F (−272°C) have been noted in connection with paramagnetic salt measurements.

Resistor sizes of 0.1 to 1 watts and ambient resistance values up to 150 ohms exhibit a large increase in resistance below −424°F (−253°C). Reproducibilities on the order of 0.2% are obtainable.

Small size, low cost, and general availability make their use attractive in cryogenic work.

The influence of stray radio interference and a loss of sensitivity are drawbacks. Variations between resistors make calibration difficult.

In addition to carbon resistors and conventional RTDs or thermistors there is a variety of special resistors used as temperature sensors. One example consists of a thin nickel film deposited onto an electrically insulating substrate in a reducing atmosphere. The sheet resistance is a function of the heat treating temperature, cycle time, and thickness. The different designs of special temperature-sensitive resistors are too numerous to mention.

CAPACITANCE CABLE FOR DETECTING HOT SPOTS

A cable consisting of an electrical conductor, a dielectric, and a conductive plastic can be used to detect hot spots. The polymer in the over-temperature locator cable increases its electrical resistance with temperature. Therefore when a hot spot evolves anywhere along the length of the cable, this will result in a drop in capacitance, as measured from one end of the cable. The location of the hot spot can be determined by comparing the measured capacitance from the control end of the cable with the capacitance of the full length of the cable.

FLUIDIC SENSORS

The fluidic sensor is a device for converting gas temperatures into gas pressure. Actually, it is a beat-frequency detector system which contains no moving parts other than the gas.

One type of design is shown in Figure 8b. The oscillator is a two-chamber resonator in which the entering gases are split by a knife edge. The gases are reflected from one chamber into the other, setting up oscillations whose frequency is proportional to the square root of the absolute temperature.

A reference signal input from a temperature-sensitive resonant oscillator is compared with the unknown in a beat-frequency detector made of beam-deflection fluid amplifiers. The frequency of its output is the beat frequency of the combined reference and oscillator signals. The components in the frequency converter create a steady pressure proportional to the beat frequency. Accuracies of about 2% and temperatures up to 2000°F (1093°C) are claimed for the system.

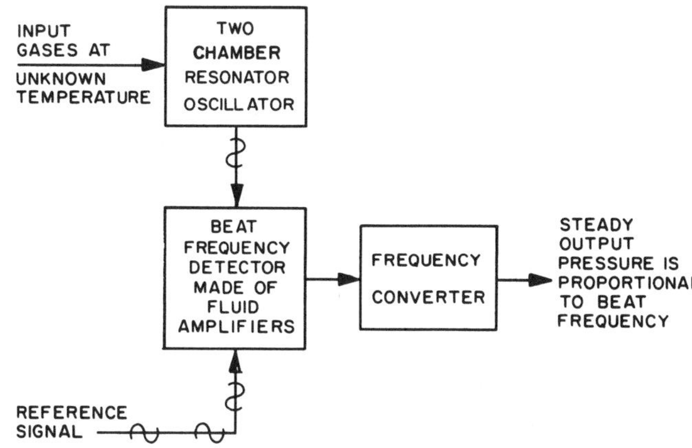

FIG. 8b
Fluidic temperature-to-pressure converter.

JOHNSON NOISE THERMOMETER

Another thermometer under development capitalizes on the phenomenon that electrical resistors provide a voltage related to thermal noise. These devices also use rhenium sensors. They are referred to as the Johnson noise detection-type thermometers.

LIQUID CRYSTALS

Used in nondestructive testing for surface temperature measurements, liquid crystals undergo a series of color changes as temperature varies. They are an organic compound which is physically liquid, but which exhibits optical properties similar to those of a crystalline solid.

A number of solutions are available, from the minimum temperature of about 68°F (20°C) to a maximum of approximately 340°F (170°C). The solutions are packaged in kit form for various ranges within these limits. Mixtures are made covering spans as narrow as 4°F (2.2°C) within the selected range.

Temperature is read by comparing the color exhibited when a thin coating is subjected to the conditions under question to a standard reference color. Response speed is less than one second. The indication is continuous and reversible. Cost is low. Disadvantages are manual preparation and limited range, in addition to lack of automatic readout.

Liquid crystals have also been used in fiber-optic thermometers. A pellet of liquid crystal is inserted at the tip of an optical fiber, and as light at different wavelengths is sent through the fiber, the reflection peaks from the liquid crystal are related to temperature.

PARAMAGNETIC SALTS

Magnetic thermometry has been developed chiefly to measure temperatures near absolute zero (below −458°F, or −272°C). The temperatures themselves are obtained by adiabatic demagnetization of a paramagnetic salt. An isothermal magnetization at the lowest attainable liquid helium

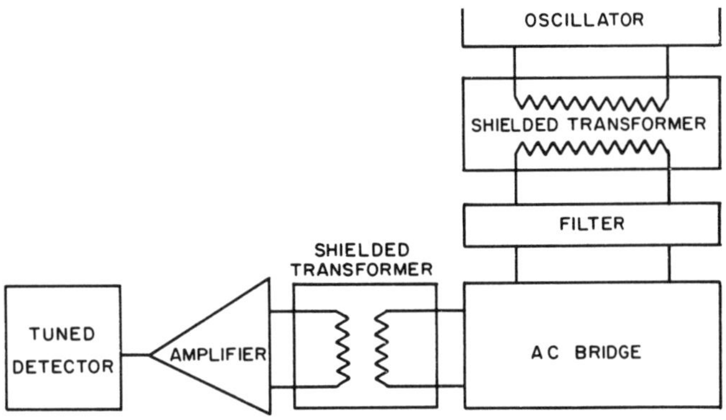

FIG. 8c
Circuitry for temperature-measurement by paramagnetic salts.

temperature (about −458°F, or −272°C) followed by an adiabatic demagnetization is used. The entropy is decreased, with a simultaneous heat flow from the sample, when the magnetic ions are oriented parallel to the field. During subsequent adiabatic demagnetizations the entropy of the salt remains constant, if demagnetization is reversible, and temperature decreases.

To obtain the temperature, some temperature-dependent quality of the salt under investigation is used, such as the magnetic susceptibility.

If a sphere or rotational ellipsoid of an isotropic paramagnetic salt is located in the homogeneous part of the magnetic field of a coil of a mutual inductance or a self-inductance, the inductance of the coil is a function of the temperature. Inductance can be measured with an AC bridge (as shown in Figure 8c) whose balance is independent of frequency. A galvanometer can be used for detection. Effective shielding is a requirement.

In a paramagnetic salt with a coil surrounding it, self-inductance is related to temperature. An Anderson AC bridge has been used to measure magnetic temperature in such a situation. The relationship between self-inductance and susceptibility of a salt has been found to be linear when the ellipsoidal or spherical salt piece is placed in the homogeneous part of the measuring field.

Accuracy of the magnetic method has been estimated on the order of .001°F (.00056°C). The method is the best available for measurements near absolute zero.

SPECTROSCOPIC TEMPERATURE MEASUREMENT

Spectroscopic methods are often used to measure the temperature of hot gases. They are, in fact, the only possible way to measure the surface temperature of stars.

The spectroscope in its simplest form is the familiar triangular glass prism which breaks up light from a hot object into its constituent colors (its spectrum). The chemical composition of glowing gas is determined from the pattern of dark (Fraunhofer) lines which appear across the spectrum.

Many procedures for temperature determination from the spectrum have been developed, such as measurement of brightness and actual color, reversal temperatures, population temperature estimates, measurements made of spectral line shifts in ionized gases, and many others. These are all laboratory techniques seldom employed industrially because of their complexity and relatively high costs. For further information on this subject the reader is referred to Chapter 56, "Spectroscopic Methods of Temperature Measurement," of the book *Temperature—Its Measurement and Control in Science and Industry,* Volume III, Part I.

THERMOGRAPHY

The strong temperature dependence of the brightness of certain luminescent materials may be converted into a pattern of color which can be recorded photographically. A thin layer of this material is placed on the surface to be investigated and is suitably excited with ultraviolet radiation in a darkened room. The temperature is implied by the brightness of the coating in comparison to the brightness of the same coat at a known temperature.

The sensitivity of the phosphors used gives a 10% brightness change per °F and this can be picked up with a relatively crude system of photometry. Temperature range for this type of measurement is from 32 to about 750°F (0 to about 400°C).

References

1. Plumb, H.H., "Temperature, Its Measure and Control in Science and Industry," 5th Symposium on Temperature, held in 1971 by NBS, AIP, and ISA.
2. Lynnworth, L.C., "Sound Ways to Measure Temperature," *Instrumentation Technology,* April 1969.

Bibliography

Bidini, G., "A Method for the Thermo-Fluid Dynamic Design of Gas Turbine Total Temperature Probes," 1991 ISA Conference, Paper #91-0511.

Bliss, P., and Morgan, R.K., "Errors in Temperature Measurement," *Instrumentation Technology,* March 1971.

Hormuth, G.A., "Ways to Measure Temperature," *Control Engineering,* Reprint No. 948, 1971.

Kerin, T.W., and Katz, E.M., "Temperature Measurement in the 1990s," *InTech,* August 1990.

Krigman, A., "Temperature Instrumentations: Recent Patent Activity," *InTech,* August 1983.

Leewis, W., "The International Temperature Scale of 1990," 1991 ISA Conference, Paper #91-0302.

Schooley, James F., "State of the Art of Instrumentation for High Temperature Thermometry," Argonne National Laboratories' 1977 Symposium, Publication No. ANL-73-7.

Shepard, R.L., et al., "Development of Long Life, High Reliability, Remotely Operated, Johnson Noise Thermometer," 1991 ISA Conference, Paper #91-0306.

9 Pneumatic and Suction Pyrometers

B. G. LIPTÁK (1993)

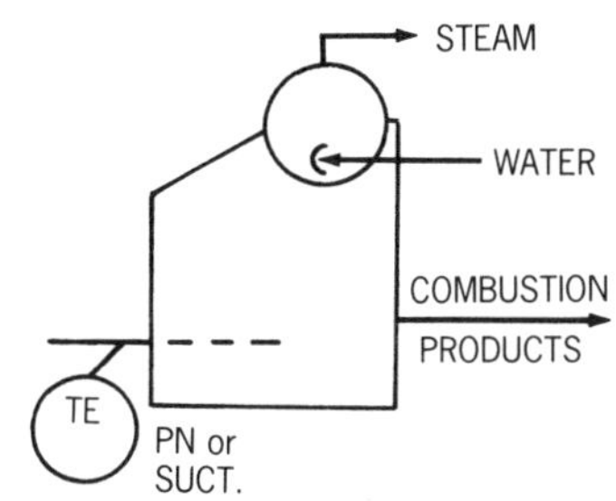

Flow Sheet Symbol

Types:	Furnace gas temperature can be measured by taking it directly to a thermocouple or by measuring its pressure drop ratio as it is aspirated through a sample probe
Applications:	Temperature profiling of boilers and furnaces during commissioning
Length of Probes:	6 to 30 ft (2 to 10 m); the permissible insertion length is a function of the maximum operating temperature
Inaccuracy:	2% of actual temperature
Response:	98% in 10 to 20 seconds
Probe:	Stainless steel water-jacketed units with 2 to 3 in. (50 to 75 mm) diameters
Temperature Range:	0 to 2900°F (0 to 1600°C) with thermocouple and 0 to 4500°F (0 to 2500°C) with pneumatic probes
Fan or ejector requirements:	The fan must generate about 20 inches of water column (0.05 bar, or 37 mmHg) vacuum and pull a flow rate of about 40 cfm (2000 lpm) of furnace gas
Cost:	$5000
Partial List of Suppliers:	Land Combustion, a Div. of Land Instruments International Inc.

The measurement of hot gas temperatures is necessary for the understanding of combustion processes. This measurement cannot be done with conventional thermometers because of interference from the radiated heat of the surroundings. When a conventional thermocouple is used in an attempt to make such a measurement, the amount of interference is a function of the relative values of the radiant heat transfer coefficient of the surroundings and the convective heat transfer coefficient of the gas. When a standard thermocouple is used in a cold furnace, the resulting error can be several hundred degrees. Pneumatic and suction pyrometers have been developed in order to lower this error.

SUCTION PYROMETERS

The suction pyrometer consists of a sheathed thermocouple (sheathed against chemical attack) located inside a radiation shield at the tip of a suction pipe (Figure 9a). The combustion gases are sucked through the shield and over the thermocouple at high velocity by aspirating equipment. The efficiency of this aspirating thermocouple is a function of the quality of the radiation shield and of the suction flow rate. If, for example, a suction pyrometer has a 100°C error without suction and a 10°C error when the suction flow is on, it is said to have an efficiency of 90%.

The suction pyrometer probe is usually made of stainless steel and is water-cooled. These probes are designed for high gas velocities of 500 fps (152 mps). At such velocities the radiation shields usually produce better than 90% pyrometer efficiencies. The furnace gases can be pulled out by fans or by air or steam ejectors. The main limitations of this design include plugging of the probe when the combustion products are dusty (as in pulverized coal burners) and temperatures exceeding 2912°F (1600°C).

PNEUMATIC PYROMETERS

The pneumatic probe was invented in 1893 for use in blast furnaces. It is superior to suction pyrometers in that dust in the combustion gases does not plug it as easily and that it can measure very high temperatures, such as in turbojet engines or rockets. This pyrometer can detect higher gas temperatures than can the aspirating thermocouple because the thermometer does not need to reach the gas temperature (Figure 9b). The pneumatic probe can handle gases heavily laden with dust because it does not have a radiation shield, which is prone to plug.

The pneumatic pyrometer measures the pressure drop of the hot combustion gases through one restriction at the probe tip and measures their pressure drop again after they have been cooled through an identical restriction (see Figure 9b). Because the mass flow rate of the gas is the same across these restrictions, the only reason these two pressure drops are different is because of the change in the gas temperature,

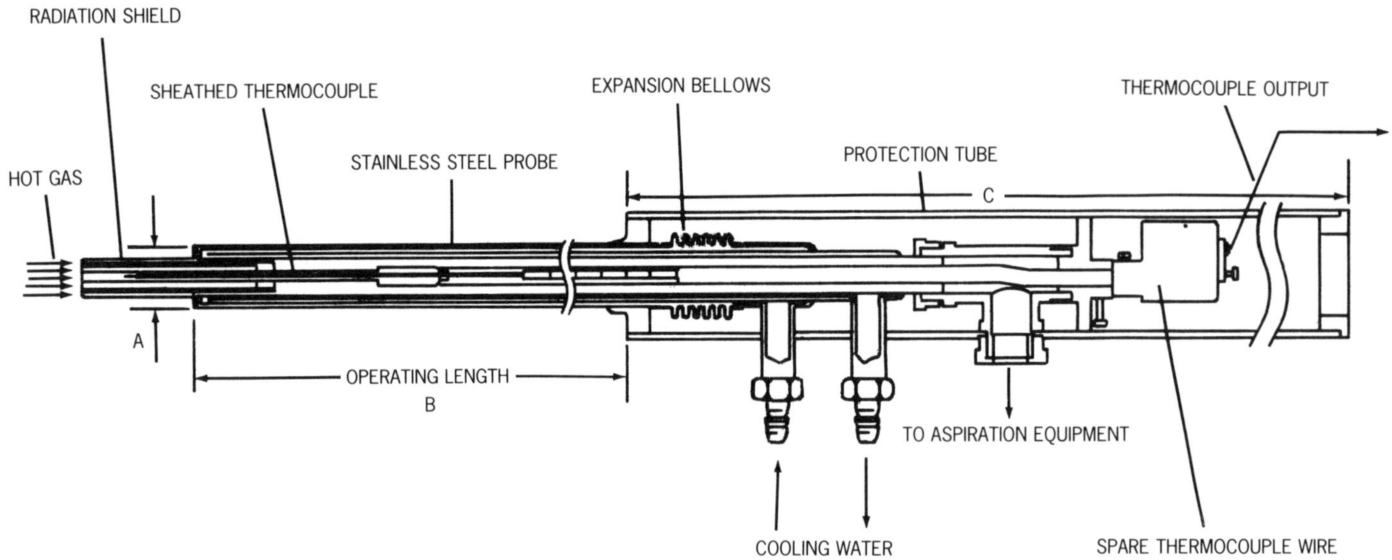

FIG. 9a
Sheathed thermocouple detects the hot gas temperature in an aspirating suction pyrometer. (Courtesy of Land)

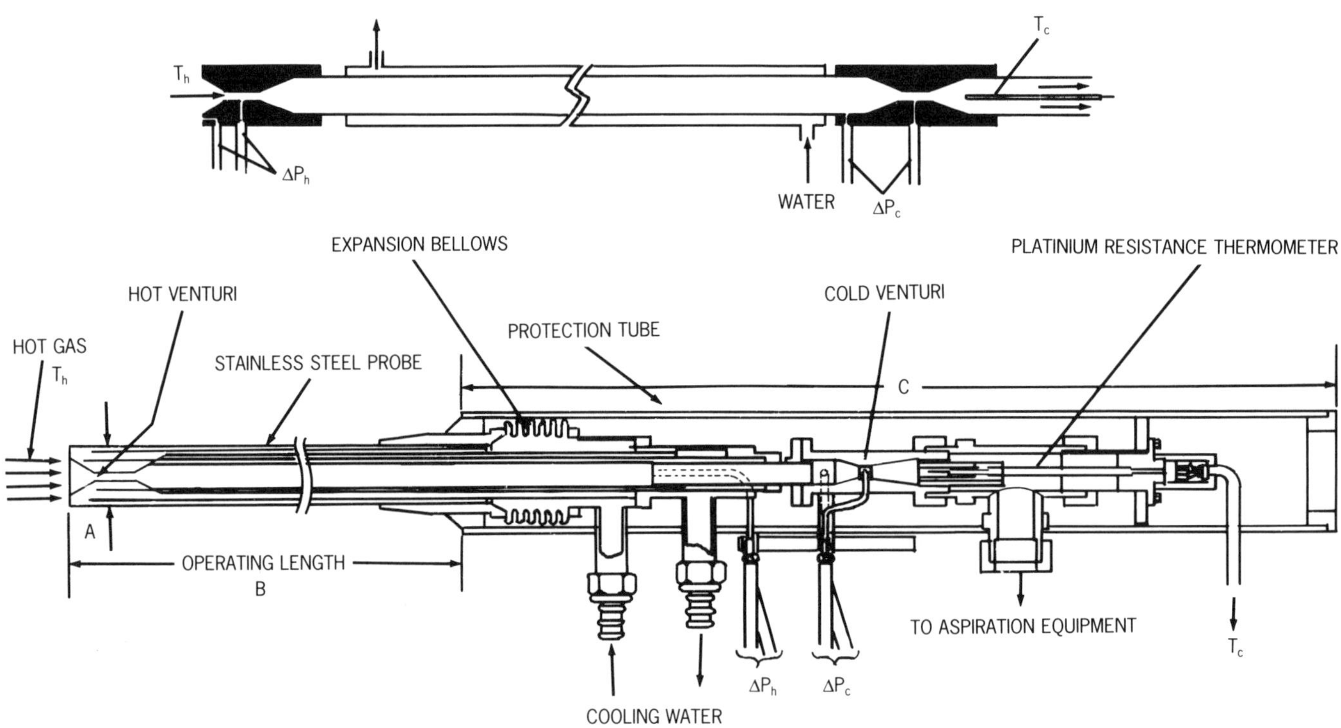

FIG. 9b
Pneumatic pyrometer determines hot gas temperature by multiplying cold end temperature (T_c) with the ratio of pressure drops across the "hot" and "cold" venturis. (Courtesy of Land Combustion, a Div. of Land Instruments International Inc.)

which affects its density. The pressure differentials across the "hot" and "cold" venturis are inversely proportional to the gas densities. Therefore they are directly proportional to the absolute temperatures at the "hot" and "cold" venturis. As the temperature at the cold venturi (T_c) is measured, the hot gas temperature (T_h) can be calculated by multiplying T_c and a configuration constant K with the ratio of P_h/P_c. Therefore, the absolute temperature of the gas is:

$$T_h = KT_c\,(P_h/P_c) \qquad \mathbf{9(1)}$$

Relative to the suction pyrometer, the pneumatic pyrometer requires more utilities and more design attention as almost all units must be uniquely designed. Both pyrometer designs are available with a full range of display and recording instruments for plotting the required temperature profiles.

Bibliography

Kerin, T.W., and Katz, E.M., "Temperature Measurement in the 1990s," *InTech,* August 1990.

Peterson, W., "Choosing the Right Temperature Sensor," *InTech,* April 1991.

Plumb, H.H., "Temperature, Its Measure and Control in Science and Industry," 5th Symposium on Temperature, held in 1971 by NBS, AIP, and ISA.

Schooley, J.F., "State of the Art of Instrumentation for High Temperature Thermometry," Argonne National Laboratories' 1977 Symposium, Publication No. ANL-73-7.

10 Pyrometric Cones

T. J. CLAGGETT, R. W. WORRALL (1969, 1982) **B. G. LIPTÁK** (1993)

Temperature Ranges:	Cones cover a range from 1100 to 3650°F (593 to 2010°C)
Costs:	$3 to $5 for a box of 50 cones; $0.50 to $1.50 per plug
Inaccuracy:	Indication is affected by speed of heat-up and by time spent at target temperature; usual error is about 5°F (3°C)
Partial List of Suppliers:	Ball Research Inc. (cone); Biosynergy Inc. (freeze-thaw indicator); Edward Orton Jr. Ceramic Foundation (cones); Electronic Development Labs Inc. (crayons); Hub Material Co. (crayons); Markal Co. (stick-ons); Shell Oil Co.; Solder Absorbing Technology Inc. (crayons, lacquers); Tempil Division of Big Three Industries Inc. (crayons, paints); Thermax USA Div. of Everest Assoc. (paints, crayons, labels)

For centuries, manually operated furnaces have been temperature-controlled by the operators' placing a heat-sensitive object inside the furnace and observing the status of that indicator through a peephole. The most easily observed physical changes were found to be changes in color or melting. These devices (color indicators) are discussed in Section 4. Pyrometric cones have been used as temperature indicators and are still used as end-point indicators in such batch processes as firing in pottery furnaces. They can be small, expendable, plugs, chips, or geometrically shaped objects whose purpose is to accompany the products through a heating cycle. The physical or metalurgical changes that occur indicate the temperature reached in the process.

THE CERAMICS INDUSTRY

The German ceramist Herman Seger invented the first pyrometric cone in 1886.[1] The individual cones look like truncated pyramids. The numbering of the different cones available on the market and the temperatures at which they melt are given in Table 10a. The cone melting temperatures listed in the table are based on a heat-up rate of 170°C per hour. If the heat-up rate is slower, the cones bend at slightly lower temperatures; if it is faster, it will take a little higher temperature to bend the cones. Because the pyrometric cone measures the effects of both temperature and the length of firing time, it acts as a heat integrator, a function that cannot be easily reproduced by thermocouple-type pyrometers. For this reason the ceramic industry continues to use cones, in addition to recording pyrometers, even in its most modern kilns.

The indicator material is generally quite similar to the substance of the work under test. Pyrometric cones are actually composed of ceramic materials very carefully blended to soften at a certain temperature. The slender cone is slightly tilted from the vertical; when its softening point is reached the tip bends over and may actually touch the base. This action can be watched through the window of the firing furnace or its condition can be studied after cooling. Observation of a fired cone will show the experienced operator if the furnace atmosphere was oxidizing, reducing, or carburizing. If the latter has taken place the cone will have formed a shell less dense than the interior. Presumably the work will have taken on the same characteristic.

The cones are set in a plaque of fireclay (called a ''cone pat'') close together and tipped at about 8 degrees from the vertical toward the cone that is expected to bend first. The cone pat is located in front of the peephole where the operator (or a closed-circuit TV camera) can observe its status. Usually three cones are placed on a pat: the controlling target cone is in the center, a cone that melts at a lower temperature is in front of it and one that melts at a higher temperature is behind it. Figure 10b illustrates the visual appearance of a cone pat throughout a batch that was fired to a #4 cone target, corresponding to 2174°F (1190°C). The softening points can be selected very accurately, and accuracies of 2 to 5°F (1 to 3°C) can be obtained.

As an alternate to cones, the indicator may take the shape of a long cylindrical bar. The bars are supported at their ends with axes horizontal. On temperature rise they soften and sag at the middle under gravity. The deformation serves as a measure of temperature.

Another group of indicators operates by shrinkage rather than deformation. After removal from the furnace the diameter of a hole in the indicator, or perhaps the indicator's

TABLE 10a
*The Numbering of Pyrometric Cones**

Cone	Centigrade	Fahrenheit	Color of Fire	Cone	Centigrade	Fahrenheit	Color of Fire
15	1435	2615	white	07	990	1814	orange
14	1400	2552		08	950	1742	
13	1350	2462		09	930	1706	
12	1335	2435		010	905	1661	
11	1325	2417		011	895	1643	cherry red
10	1305	2381		012	875	1607	
9	1285	2345		013	860	1580	
8	1260	2300		014	830	1526	dull red
7	1250	2282		015	805	1481	
6	1230	2246		016	795	1463	
5	1205	2201	yellow	017	770	1418	
4	1190	2174		018	720	1328	
3	1170	2138		019	660	1220	
2	1165	2129		020	650	1202	
1	1160	2120		021	615	1139	
01	1145	2093		022	605	1121	
02	1125	2057					
03	1115	2039					
04	1060	1940					
05	1040	1904					
06	1015	1859					

*From Reference 2.

length, is measured and compared with the original dimension.

Like color indicators, pyrometric ceramics should not be considered exact temperature measuring devices. The fusion, bending, and/or shrinking that they undergo is a *time-temperature* relationship and, as such, it is only useful to determine the end point of the specific job. This property is frequently more important than an exact measurement of the instantaneous temperature. The use of this type of indicator may almost be considered an art.

While pyrometer cones are not well suited to automatic process control, they are inexpensive and valuable quality control tools in guaranteeing repeatable qualities of ceramic and similar batch products from kilns and furnaces.

ENGINE TEST RESEARCH

An entirely different material, used in a similar manner, is the metal test plug. This small device can tell temperature by a change in hardness that results from the heat treatment it has received. One use is to have it located carefully in an operating engine, in an otherwise inaccessible spot, where it will respond to the temperatures that occur during operation. When the test is over the plug is removed and carefully

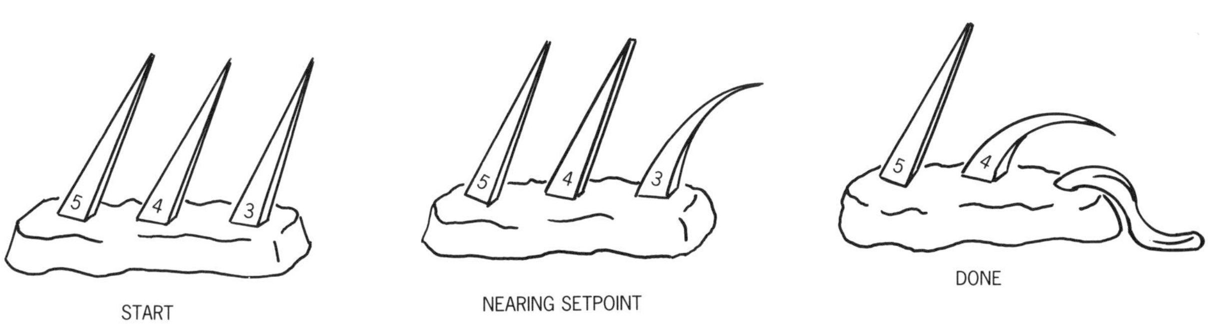

FIG. 10b
This is the appearance of the cones at different times during a batch which has a target temperature set by Cone #4.[2]

analyzed to determine the change in hardness along the horizontal axis.

Time is again a factor, but metal responds much faster than ceramic material. Exposures of less than one second duration can be detected.

Advantages of this class of temperature sensors are their relative economy and design for a very specific job. Their shortcomings are obvious.

References

1. Rhodes, D., *Kilns Design, Construction and Operation,* Radnor, Pennsylvania: Chilton Book Co., 1968.
2. Kenny, J.B., *Pottery Making,* Radnor, Pennsylvania: Chilton Book Co., 1974. "Temperature-Sensitive Paints, Crayons," *Measurements and Control,* December 1991.

Bibliography

Considine, D.M., *Encyclopedia of Instrumentation and Control,* New York: McGraw-Hill, 1971.

Hormuth, G.A., "Ways to Measure Temperature," *Control Engineering,* Reprint No. 948, 1971.

Plumb, H.H., "Temperature, Its Measure and Control in Science and Industry," 5th Symposium on Temperature, held in 1971 by NBS, AIP, and ISA.

"Temperature-Sensitive Paints, Crayons," *Measurements and Control,* December 1991.

11 Radiation and Infrared Pyrometers

T. J. CLAGGETT, R. W. WORRALL (1969, 1982) **B. G. LIPTÁK** (1993)

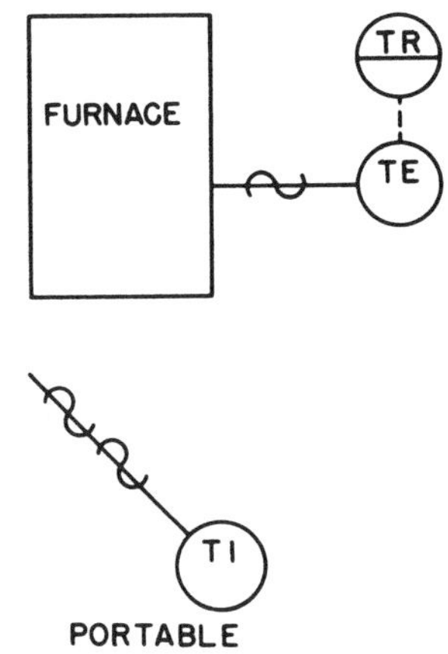

Flow Sheet Symbol

Types:
A. Optical (brightness)
B. Ratio (two-color)
C. Total (wide-band)
D. Narrow-band
Note: Heat-flow and fiber-optic designs are covered in other sections (Sections 1 and 5, respectively)

Applications: Hot metal working, plastic film processing, and glass processing. Applications are not limited to noncontact high-temperature processes. Low-temperature applications include infrared photographs of accident victims to locate internal bleeding, aerial photographs to detect pollution of water bodies or the need for crop irrigation, and many others

Wavelengths Used (Microns): Inexpensive total radiation detectors cover 0.7 to 8. For penetrating intervening atmosphere, 2.2 to 3.8; for metals, a narrow band around 2.2; for flames, 4.5; for glass, below 8

Temperature Ranges: 1400 to 6300°F (760 to 3500°C) for optical; −40 to 7000°F (−40 to 4000°C) for narrow-band and total radiation pyrometers

Distance and Target Size: Maximum distance (D) to target is about 100 ft (30 m); minimum target diameter (d) is about 0.05 in. (1.5 mm). The field of view of the pyrometer is its D/d ratio and varies from 3:1 to 900:1. Single-wavelength units detect average temperature of field of view; dual-wavelength units detect the hottest part of target within the field of view

Linearity: Nonlinear, most vary with the 4th power of absolute temperature

Main Sources of Errors: Variations in target and background emittance; aging of radiation source; background interference; focusing and targets too small to fill the view; radiation absorption by dust, smoke, moisture, atmosphere, or dirt on windows

Methods of Focusing: Most units use lenses to focus on the target. (Fiber-optic designs are discussed in Section 5.)

Inaccuracy:
A and B. 1 to 2% full scale (FS), 0.5% of reading on narrow spans
C. 0.5 to 1% FS
D. 0.25 to 2% FS
It should be noted that under laboratory conditions the precision can reach 0.1°C (0.2°F).

Costs: Portable general purpose infrared thermometer units start at $400 and can reach over $3000 in optical (A) designs. Permanently installed general-purpose, two-wire industrial transmitters start at about $1500. These can be wide- or narrow-band (C and D) units. Ratio-type transmitters (B) start at around $4000. The optical (A) transmitters cost over $5000 and are not widely used. When installing pyrometers on a furnace or on a boiler, accessories such as cooling jackets, air purge assemblies, safety shutters, and sighting or target tubes are also needed and can more than double the cost of the pyrometer itself. Additional costs involve testing, maintenance, and periodic calibration. When all costs are added up the total cost is higher than that of any other thermometer.

Partial List of Suppliers: Accufiber Inc.; AEMC Corp.; Agema Infrared Systems; ANASCO Corp.; Barber-Colman Industries Instruments Div.; Barnes Engineering Co.; Capintec Instruments Inc.; Chino Works America Inc.; CI Systems Inc.; Concept Engineering; Dowty Custom Electronics Inc.; E2 Technology Corp.; Elan Technical Corp.; Elcontrol USA Inc.; Electronic Development Labs; Enercorp Instruments Ltd.; Epic Inc.; The Eppley Laboratory; Everest Interscience Corp.; Exergen Corp.; Extech International Corp.; Gentri Controls Inc.; C.S. Gordon Co.; Heimann Systems, Siemens Components; Honeywell Industrial Automation; Horiba Instruments Inc.; IKE (Germany); International Thermal Instruments; Ircon Inc.; IR-Onics Inc.; Intronics Systems Inc.; Kane-May Measuring Instruments; Kobold Instruments, Inc.; Land Infrared Inc.; Laser Precision Corp.; Leeds & Northrup Co.; LFE Corp.; Linear Corp.; Luxtron Corp.; Material Control Inc.;

M.C. Products; Mikron Instrument Co.; Nanmac Corp.; Omega Engineering Inc.; Pyrometer Instrument Co.; Quantum Logic Corp.; Ratio Controls Corp.; Raytek Inc.; Rexanne Products; Rosemount Analytical Inc.; Siemens Allis Inc.; Simpson Electric Co.; S-Products Inc.; Square D Infrared Measurement; Suscon International Inc.; Telatemp Corp.; Testoterm Inc.; Thermometrics Inc.; Trans-Met Engineering Inc.; TRW Dynalco Controls Div.; Universal Enterprises Inc.; UTI Instruments Co.; Vanzetti Systems Inc.; Wahl Instruments Inc.; Watlow Infrared; Weschler Instruments; Williamson Corp. *Note*: Most popular are Ircon and Williamson.

It was only a few hundred years ago that physicists abandoned the view that heat is a substance and accepted that it is a form of energy which is transferable from one material to another by either conduction, convection, or radiation. Since that time it was discovered that all materials, at all temperatures, down to near absolute zero, radiate electromagnetic energy. All electromagnetic radiation travels at the speed of light (186,000 miles per second). It was also learned that the wavelength at which this radiation occurs depends on the temperature of the material. The temperature of the material determines both the quantity and the type of energy radiated. As the temperature rises, the wavelength of radiation drops and its frequency rises (Figure 11a). For example at −100°C, the range of radiant energy wavelengths is between 10 and 100 microns, while at 1000°C they are between 1 and 10 microns. As the temperature rises, the dominant form of heat radiation shifts toward the shorter wavelengths; infrared (2 to 20 microns), near infrared (0.7 to 2 microns), visible (0.4 to 0.7 microns), ultraviolet (0.04 to 0.4 microns); at extremely high temperatures the hot objects will also radiate X-rays and gamma rays. In industrial applications the bulk of thermal radiation occurs in the infrared range, and therefore these thermometers are often referred to as infrared (IR) thermometers.

This section is started with the definition of terms used in radiation pyrometry. This series of terms is followed by a discussion of some theoretical aspects of pyrometry, including such topics as emissivity. The description of the specific thermometer designs, features, and installations concludes the section.

DEFINITION OF TERMS

ABSORBANCE (A) Ratio of radiant energy absorbed by a body to the corresponding absorption of a blackbody at

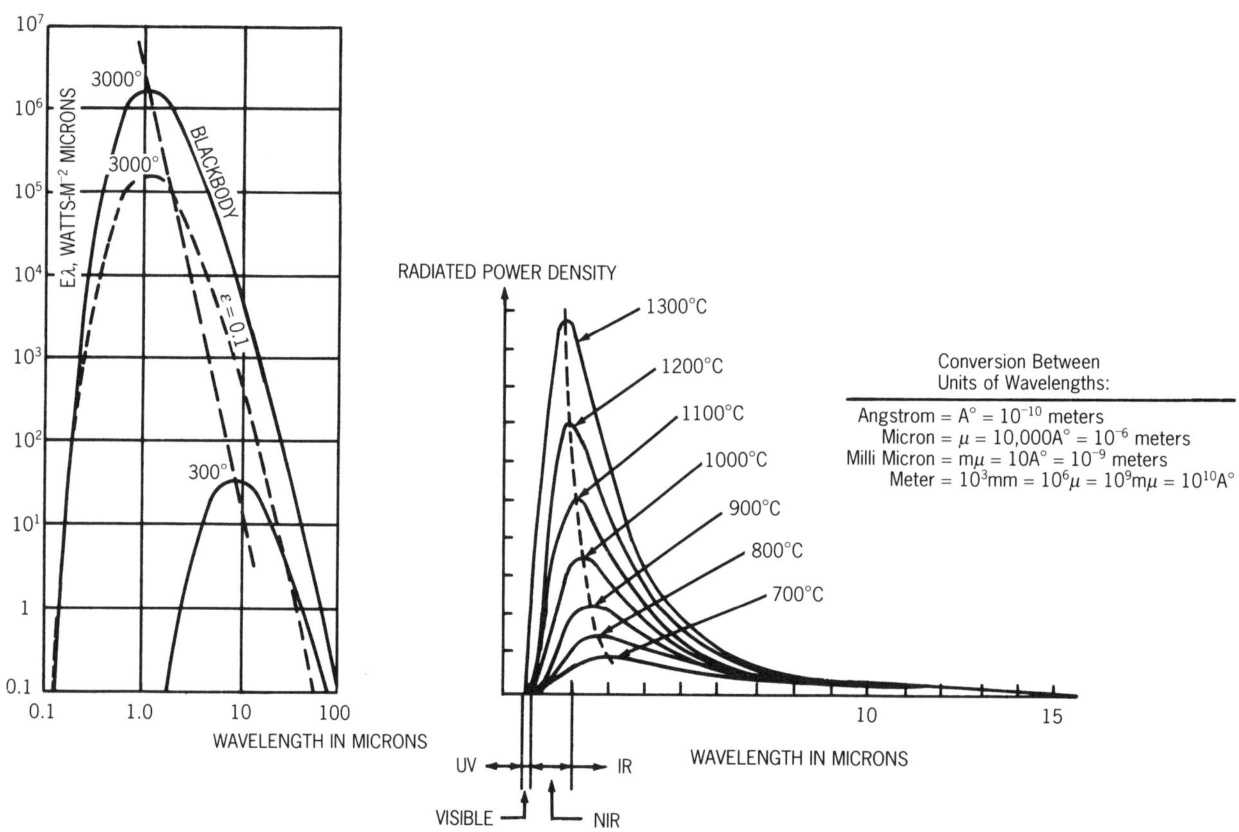

FIG. 11a
Blackbody radiation as a function of temperature. Dotted line on the left describes a gray body having an emissivity of 0.1.

the same temperature. Absorbance equals emittance on bodies whose temperature is not changing. (A = 1 − R − T, where R is the reflectance and T is the transmittance.)

BAND PASS FILTER An optical or detector filter which permits the passage of a narrow band of the total spectrum. It excludes or is opaque to all other wavelengths.

BLACKBODY The perfect absorber of all radiant energy that strikes it. The blackbody is also a perfect emitter. Therefore, both its absorbance (A) and emissivity (E) are unity. The blackbody radiates energy in predictable spectral distributions and intensities which are a function of the blackbody's absolute temperature (Figure 11a). A blackbody can be configured as shown in Figure 11b.

BOLOMETER Thermal detector which changes its electrical resistance as a function of the radiant energy striking it.

BRIGHTNESS PYROMETER Uses the radiant energy on each side of a fixed wavelength of the spectrum. This band is quite narrow and usually centered at 0.65 microns in the orange-red area of the visible spectrum.

DETECTOR A device that measures the amount of energy radiated by an object. Can be a thermal detector or a photodetector. Thermal detectors respond to radiation by changing their volume, capacitance, or generation of millivoltages; they can be thermocouples, thermopiles, pneumatic detectors, or bolometers. Their common feature is their relatively slow response. Photodetectors are semiconductors which produce a signal in proportion to the photon flux which strikes them.

EMISSIVITY OR EMITTANCE (E) The emissivity of an object is the ratio of radiant energy emitted by that object divided by the radiant energy which a blackbody would emit at that same temperature. If the emittance is the same at all wavelengths, the object is called a gray body. Some industrial materials change their emissivity with temperature and sometimes with other variables also. Emissivity always equals absorption and it also equals 1 minus the sum of reflectance and transmittance (E = A = 1 − T − R).

GRAY BODY This is an object having an emittance of less than unity, but this emittance is constant at all wavelengths (over that part of the spectrum where the measurement takes place). This means that gray-body radiation curves are identical to the ones shown in Figure 11a, except that they are dropped down on the radiated power density scale.

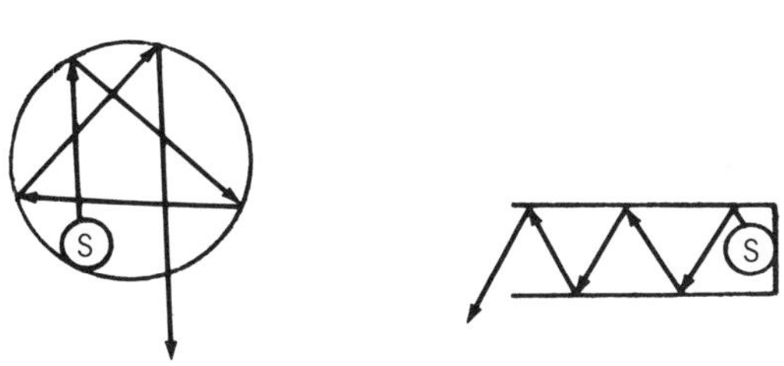

FIG. 11b
Emissivity of blackbody sources in the laboratory is around 0.99.

INFRARED That portion of the spectrum whose wavelength is longer than that of red light. Only the portion between 0.7 and 20 microns gives usable energy for radiation detectors.

MECHANICAL EMISSIVITY ENHANCEMENT Mechanically increasing the emissivity of a surface to near-blackbody conditions (using multiple reflection).

MICRON .001 millimeters. 10,000 Angstrom units. A unit used to measure wavelengths of radiant energy.

NARROW-BAND PYROMETER A radiation pyrometer that is sensitive to only a narrow segment of wavelengths within the total radiation spectrum. Optical pyrometers are one of the devices in this category.

OPTICAL PYROMETER Also called brightness pyrometer, it uses a narrow band of radiation within the visible range (0.4 to 0.7 microns) to measure temperature by color matching and other techniques.

PHOTODETECTOR Measures thermal radiation by producing an output through release of electrical changes within its body. They are small flakes of crystalline materials such as CdS or InSb which respond to different portions of the spectrum, consequently showing great selectivity in the wavelengths at which they operate.

RATIO PYROMETER See *two-color pyrometer.*

REFLECTANCE OR REFLECTIVITY (R) The percentage of the total radiation falling on a body which is directly reflected without entry. Reflectance is zero for a blackbody, and nearly 100 percent for a highly polished surface. (R = 1 − A − T, where A is the absorbance and T is the transmissivity.)

SPECTRAL EMISSIVITY The ratio of emittance at a specific wavelength or very narrow band to that of a blackbody at same temperature.

THERMOPILE Measures thermal radiation by absorption to become hotter than its surroundings. It is a number of small thermocouples arranged like the spokes of a wheel with the hot junction at the hub. The thermocouples are connected in series and the output is based on the difference between the hot and cold junctions.

TOTAL EMISSIVITY The ratio of the integrated value of all spectral emittances to that of a blackbody.

TRANSMITTANCE OR TRANSMISSIVITY (T) The percentage of the total radiant energy falling on a body, which passes directly through it without being absorbed. Transmittance is zero for a blackbody and nearly 100 percent for a material like glass in the visible spectrum region. (T = 1 − A − R, where A is the absorbance and R is the reflectance.)

TWO-COLOR PYROMETER Measures temperature as a function of the radiation ratio emitted around two narrow wavelength bands. Also called ratio pyrometer.

WIDE-BAND (TOTAL) PYROMETER A radiation thermometer that measures the total power density emitted by the material of interest over a wide range of wavelengths.

THEORETICAL RELATIONSHIPS

The development of radiation pyrometry stems from Planck's quantum theory, developed around 1900, and from Stefan-Boltzmann's law for total radiated energy.

According to the Stefan-Boltzmann equation the total radiant power density emitted (W) by an object is directly related to the emissivity of that object (E) multiplied by a constant (SB) and by the 4th power of absolute temperature (T):

$$W = (E)(SB)T^4 \quad \textbf{11(1)}$$

where

- W is in watts/square centimeter
- E is a fraction (unity for a blackbody)
- SB is the Stefan-Boltzmann constant, having a value of 5.67×10^{-12} watts $cm^{-2}K^{-4}$
- T is absolute temperature of the object in degrees Kelvin

Planck's law of radiation goes one step beyond Equation 11(1) and predicts the level of radiation emitted per unit surface area of a blackbody at each specific wavelength. Planck's equation is rather complex and therefore is not used much in the everyday work of instrument engineers. According to Planck's law, the radiation emission peaks at shorter and shorter wavelengths as the temperature rises. This is why one can estimate the temperature of a hot iron in a fire (the shorter the wavelength emitted the whiter and therefore the hotter it is) and why the dotted line connecting the peaks at different temperatures is leaning to the left in Figure 11a. This shift in peak values is expressed by Wien's law of displacement, relating the wavelength at maximum radiation (W_l) to absolute temperature (T) and Wien's constant (Cw) as follows:

$$W_l = C_w/T \quad \textbf{11(2)}$$

where

- W_l is in microns
- C_w is in units of (microns)(degree K) and has a value of 2897

Based on the above laws, one can build thermometers by measuring either the radiation at specific wavelengths (optical and narrow-band) or all the radiation (total or wide-band). Changes in process material emissivity, radiation from other sources, and losses in radiation due to dirt, dust, smoke, or atmospheric absorption can introduce errors into the total- or single-wavelength pyrometers. For these reasons many users decide to use the more expensive ratio pyrometers.

Thermal radiation is a universal property of matter which is absent only when the material is an inert gas like helium, or is at absolute zero temperature. From this property of matter, the technique of radiation pyrometry has been developed so that it is possible to infer the temperature of an object without contacting it. This is done with an optical system which collects some or all of the visible and/or infrared energy, sometimes through a filter, and focuses it on a detector (see Figure 11c). The detector converts the concentrated energy to an electrical signal, a complex function of the absolute temperature.

Only the energy given off between about 0.3 and 20 microns is of sufficient magnitude to be of any consequence. This encompasses the visible spectrum (0.35 to 0.75 microns) and the near infrared. The intensity and distribution of this energy from a substance may be compared with that of a blackbody which radiates its energy in a theoretically predictable spectral distribution and intensity, such as that shown for several temperatures in Figure 11a. The area under the curve represents the total amount of power radiated at all wavelengths.

Real targets, however, always deviate from an idealized blackbody to some degree. The ratio of energy radiated by a real body to that of a blackbody under similar conditions is termed the emittance (E). Two other optical ratio characteristics of targets are reflectance (R) and transmittance (T) and for a body whose temperature is constant, $E + R + T = 1$ for any wavelength. The radiation pyrometer would gather the radiation from all three of these energy sources over the wavelength band to which it is sensitive. In Figure 11d, if the purpose were to measure the temperature of Object B, and if A were the same temperature as B, B would absorb, emit, reflect, and transmit radiant energy and would appear to be a blackbody.

But frequently A is not uniform in temperature nor does it completely surround B. Furthermore, B might be cooler than A or have a high reflectance which causes it to reflect extraneous sources of radiant energy. If any of these conditions prevail, the measurement of the total energy radiated by B cannot be converted exactly into temperature with the Stefan-Boltzmann law.

For best results, emittance should be high and reflectance low. Transmittance of most solid objects (with the exception of glass) will be near zero. If the process material is not solid, the radiant energy detector will actually "see" beneath the surface, or if the object is thin, right through it.

Emittance, reflectance, and transmittance are not easy factors to derive and they vary considerably with wavelength. Materials like ferrous metals with a matte surface have a high emittance at the short end of the spectrum but become lesser emitters for the longer waves. On the other hand, glass acts in nearly the opposite way, being practically

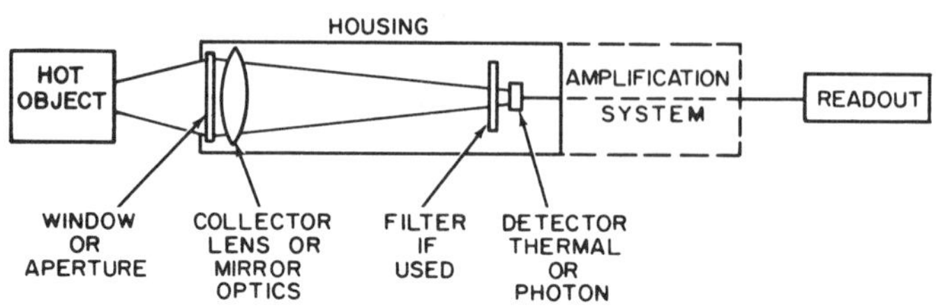

FIG. 11c
General radiation pyrometer system.

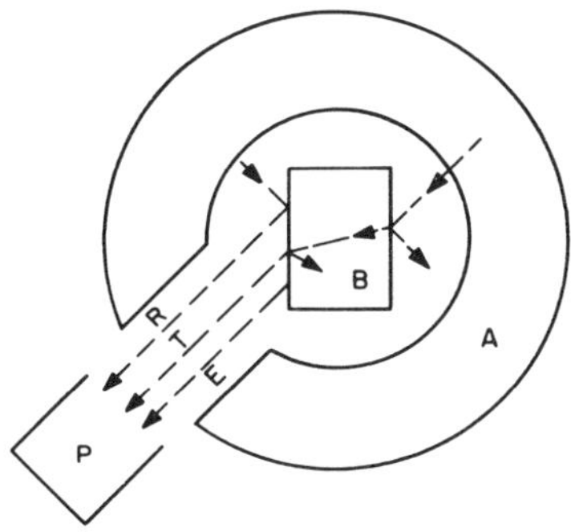

FIG. 11d
Emission, reflection, and transmission.

transparent to visible energy and almost opaque to wavelengths in the 5 to 7 micron region.

SELECTING THE RADIATION PYROMETER

In selecting the appropriate type of radiation pyrometer to be used in detecting the temperature of a particular process, one would prefer to select a wavelength-band in which the transmittance (T) is near zero (the material is opaque). In addition, one would like to select a wavelength-band that will not be absorbed by the atmosphere the radiation has to pass through. These two considerations result in a requirement for matching the pyrometer wavelength to the particular process. Figure 11e illustrates three different processes showing both the atmospheric absorption bands and the transmittance curves of the particular process materials. The arrow with the letter P for pyrometer indicates the appropriate wavelength-band selection for the radiation thermometer for each process. For polyethylene it is around 3.4 microns, for glass it is around 5 microns, for polyester it is around 8 microns.

Because of the different needs of the various processes, each radiation pyrometer manufacturer has developed a range of products operating in a variety of wavelength-bands, as shown in Figure 11f. Of the eight bands shown on this illustration, #6 can be considered for use on a polyethylene and #8 on a glass temperature measurement application. (If the glass is thick enough it can appear opaque even in the near-infrared range.)

Short wavelength units like #1, #2, and #3 can be used in iron and steel processing. Long or broad wavelength pyrometers (#7) are useful at lower temperatures where short wavelength radiation can be undetectable. Materials

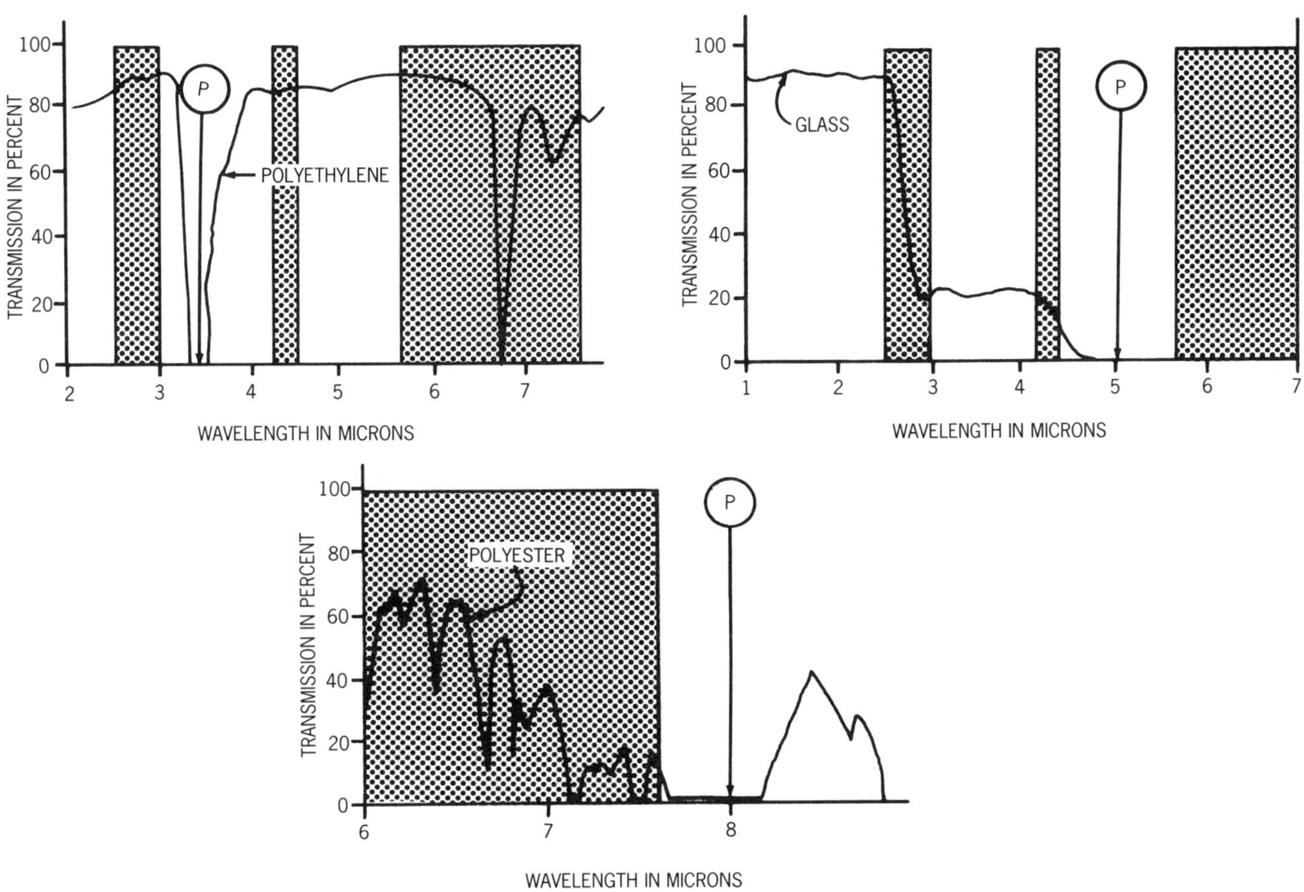

FIG. 11e
Radiation pyrometers must be selected on the basis of the transmittance of the process material and the location of the atmospheric absorption bands. (Courtesy of Ircon Inc.)

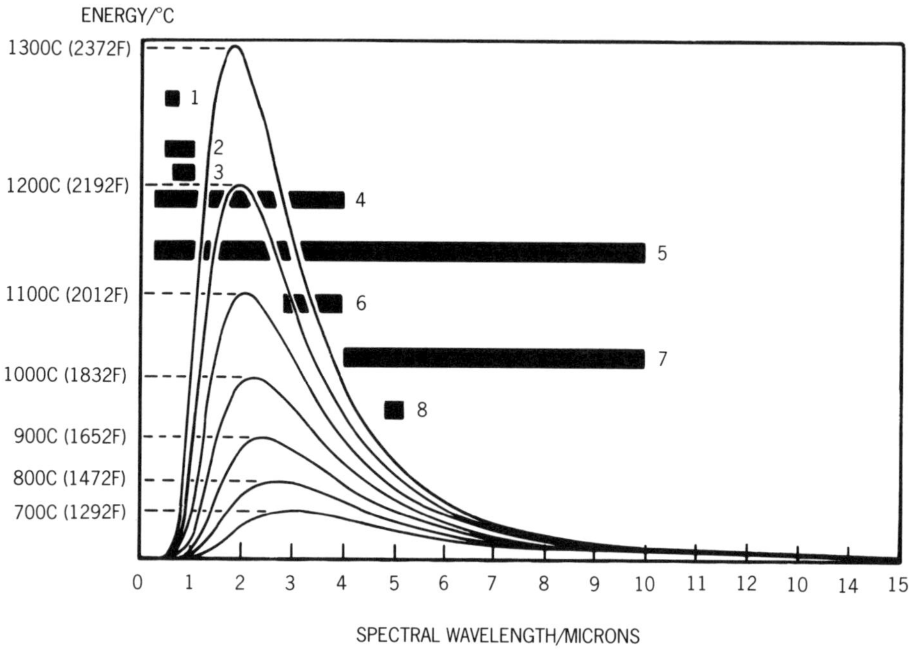

FIG. 11f
Pyrometers and their wavelength-bands marketed by a particular supplier. (Courtesy of Leeds & Northrop Co.)

like glass are transparent to shorter wavelengths (Figure 11e) and therefore shortwave pyrometers would not detect the surface temperature, but would measure the temperature below the surface. In some processes like lamp sealing this is actually desirable, and therefore the 3 to 4 micron band is used.

EMITTANCE, EMISSIVITY

The amount of thermal radiation leaving an object depends on the temperature and on the emittance of that object. If the object is a perfect emitter (a blackbody), its emittance is unity. The emissivities of almost all substances are known (Table 11i), but unfortunately the *emissivity* determined under laboratory conditions seldom agrees completely with the *actual emittance* under actual operating conditions. This is because emissivity is only one component in determining the emittance of an object; other factors (having effects that cannot be found in tables like Table 11i) include shape, oxidation, or surface finish. For this reason one is likely to use published emissivity data only when the values are near unity, as is the case for some metal oxides, ceramics, bricks, and glasses.

The uncertainties concerning emittance can be reduced by creating blackbody conditions (target tubes or target holes, which are discussed later) or by using short-wavelength or ratio pyrometers. The use of short wavelengths is useful because the signal gain is high in this region. This can be seen in Figure 11a, where the change in the amount of energy radiated per unit temperature change gets much higher when the wavelength is short. This high output response tends to swamp the effects of emittance variations and makes the performance of the short-wavelength pyrometers acceptable in such applications as iron and steel processes. The high gain of the radiated energy also tends to swamp the absorption effects of steam, dust, or water spray.

The ratio, or two-color, pyrometer overcomes emittance uncertainties but has other limitations, which will be discussed later. The emissivities of most materials, including metals, vary a great deal, and some emissivities also vary with temperature. When this is the case, the ratio thermometer is also disabled and the only solution is the actual, on-line measurement of emissivity (using laser devices, which are discussed later). The general rule is to measure the temperature of an object in a location where its emissivity is the higher and/or where the change in radiated energy is more sensitive to temperature changes than to changes in emissivity.

RADIATION PYROMETER DESIGNS

Pyrometers can be portable or permanently installed; the radiated energy can reach them through focusing on lenses or through optical fibers. In addition, they can be categorized according to the number and width of the wavelength-bands used. When the full spectrum of 0.3 to 15 microns is utilized (#5 on Figure 11f), the device is called a wide-band or total radiation pyrometer. When a single small segment of the spectrum is used (for example, #8 on Figure 11f), the design is called a narrow-band pyrometer. If that narrow band falls within the visible spectrum (#1 on Figure 11f), this subgroup of the narrow-band pyrometers is called an optical, color, or brightness pyrometer. When instead of measuring the radiation intensity of a single wavelength-band the instrument detects the ratio of the intensities of two wavelengths, the pyrometer is called a ratio, or two-color,

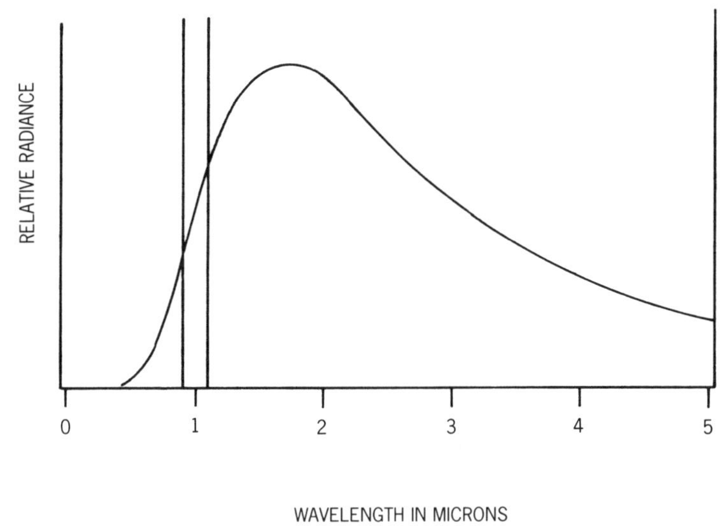

FIG. 11g
Ratio pyrometer used to detect the temperatures of molten metals, kilns, and furnaces. (Courtesy of Ircon Inc.)

thermometer (Figure 11g). When a pyrometer operates in the infrared spectrum (0.7 to about 15 microns), sometimes the unit is referred to as an infrared (IR) thermometer.

Of these many design variations, the fiber-optic units will not be discussed here, as a separate section (Section 5) is devoted to them. The others are discussed below. With the passage of time, the use of total radiation pyrometers has dropped and the use of the more accurate (and more expensive) ratio pyrometers has increased. Optical pyrometers are most often used as portable units.

Total Radiation Pyrometers

Broad-band (total) radiation pyrometers attempt to measure as much of the radiant energy coming from a hot body as possible. The simplest of the three types (total, narrow, ratio), it has substantially no selectivity for specific wavelengths other than a cutoff inherent in the optical system and is dependent on the total emittance of the surface being measured.

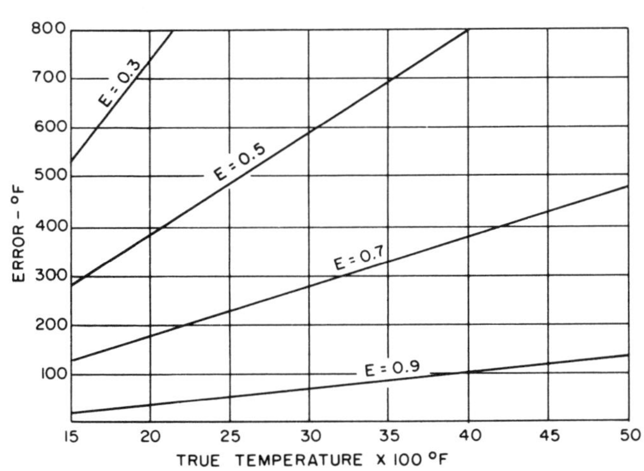

FIG. 11h
Error in reading of broad-band radiation pyrometer as function of temperature and emissivity.

Figure 11h shows the error in reading for various emissivities and temperatures when a typical instrument of this type is calibrated to read blackbody temperatures. A calibration control on the readout instrument permits the user to compensate for this error, which corrects the reading as long as the emittance does not change. The proper setting of the adjustment is made with the aid of a second thermometer known to be correct; or, knowing the nature of the target and referring to a table giving total emissivities (see Table 11i), the adjustment can be set to the correct figure.

Broad-band radiation pyrometers must have an unobstructed sighting path to the target. The presence of smoke or carbon dioxide will absorb some of the radiant energy and cause a low reading. The optical system must be kept clean, with a sighting window protected from any corrosive gases

TABLE 11i
Total Emissivities of Various Substances

Material	*Emissivity*	*Material*	*Emissivity*
Aluminum		Iron	
Unoxidized	0.06	Oxidized	0.89
Oxidized	0.19	Rusted	0.65
Brass (oxidized)	0.60	Lead (oxidized)	0.63
Calorized Copper	0.26	Monel (oxidized)	0.43
Calorized Copper		Nickel	
(oxidized)	0.19	Bright	0.12
Calorized Steel		Oxidized	0.85
(oxidized)	0.57	Silica Brick	0.85
Carbon	0.79	Silver (unoxidized)	0.03
Cast Iron		Steel (oxidized)	0.79
Oxidized	0.78	Steel Plate (rough)	0.97
Strongly oxidized	0.95	Tungsten (unoxidized)	0.07
Copper (oxidized)	0.60	Wrought Iron	
Fire Brick	0.75	(dull oxidized)	0.94
Gold	0.03		

which otherwise would come in contact with it. In furnaces, a closed-end sighting tube is frequently used to protect the optics and to provide a clear sighting path (Figure 11t). The back end of the tube assumes the same temperature as the furnace. Water or air cooling is sometimes used to keep the housing temperature cool on these installations.

Ambient temperature compensation is required for those total radiation pyrometers using a thermopile detector. Nickel wire, having a temperature coefficient opposite to that of the thermopile wires, is most often utilized. For low-temperature work, a thermostatically controlled housing is often employed to eliminate any ambient temperature fluctuations.

Narrow-Band Pyrometers

Single-band pass (narrow-band) pyrometers operate over a selected, usually narrow, band of the energy spectrum centered at a desired point. For high-temperature measurement of metals, for instance, the band might be very narrow at the 0.65 micron point, the red end of the visible spectrum where metal emissivity is highest. At this visible point the instrument might be referred to as a brightness pyrometer.

To measure gas temperatures, a band around 4.3 microns to pick up carbon dioxide might be chosen, while glass surface temperatures, previously mentioned, would be measured in the 5 to 7 micron wavelength.

Emissivity over a narrow wavelength band will not vary as much as it would over the total spectrum, but the limited-band pyrometer will suffer somewhat from a lack of sensitivity because of the reduced energy available. The use of more sensitive photon detectors helps make up this deficiency as well as providing a desired band pass. Because the energy content of the radiation increases as the wavelength gets shorter, the narrower bands are most often used to detect higher temperatures (above 500°C, or 932°F). A narrow band is usually less than 0.5 micron wide.

Ratio Pyrometers

The ratio, or two-color, pyrometer measures the energy it receives from two rather narrow bands and divides one by the other. When the two bands are chosen so that there is very little change in emissivity from one to the other, such as would be the case if the bands were close together (Figure 11g), the emissivity factor nearly cancels out. Thus, low emissivity bodies which create the most error for the broad- and single-band pyrometers do not have such a large effect on the ratio-type instrument.

The selection of the two wavelengths is not arbitrary but, like the single-band pyrometers, is made for the particular application.

Ratio pyrometers are more complex and therefore also more expensive than single-band pyrometers, but they provide better accuracies when the target behaves like a gray object or when its emissivity is a variable. (The emissivity of many substances varies with temperature.) In addition to being independent of emissivity, measurements made by ratio pyrometers are also unaffected by dust or other contaminants in the field of view (such as dirty windows), by changes in target size, or by periodic blockage of the sight path.

In some more advanced designs more than two colors (wavelengths) are used and their evaluation algorithm is more complex than a simple ratio.

Figure 11j illustrates the operation of the ratio (two-color) pyrometer where the wavelengths are alternatively selected by a rotating filter wheel. These eliminate emittance errors if the emittance is the same at both wavelengths. Unfortunately this is seldom the case; usually emittance is wavelength-dependent. Therefore, the only solution is to bring the two wavelengths very close together, but then they become insensitive to temperature. For these reasons the ratio pyrometers are not as widely used as one might expect. On the other hand they are still very useful on applications where the target is too small to fill the field of view (for example, in the drawing and coating of wires) or where the target is obscured by cool dust (as in cement kilns).

Manual Optical Pyrometers

Optical pyrometers are narrow-band or two-color radiation pyrometers which operate in the visible spectrum around the 0.65 micron point. The human eye, acting as the detector in the manually balanced type, compares a source of known radiant energy generated within the instrument by a cali-

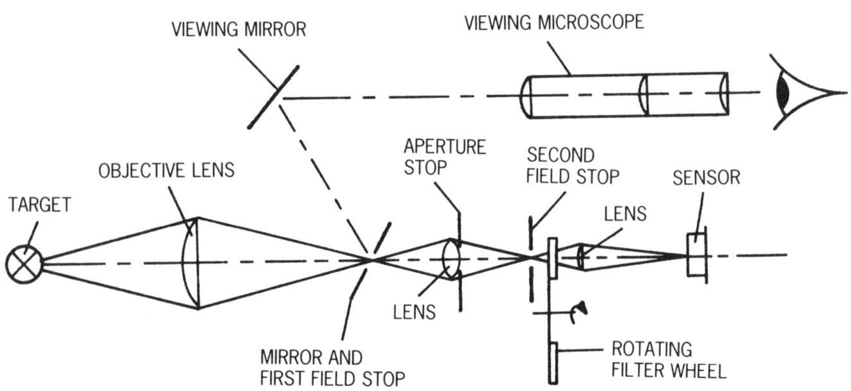

FIG. 11j
Ratio pyrometer.[1]

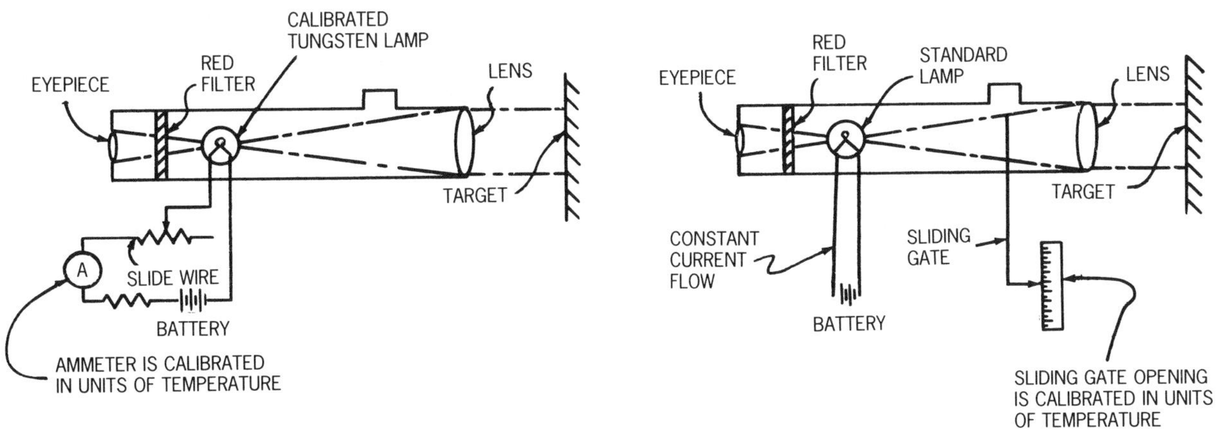

FIG. 11k
In optical pyrometers the target radiation is focused on a lamp and is compared to the radiation intensity of the light filament. Matching of the two intensities is determined visually by either adjusting the current flow through the calibrated tungsten lamp (left) or by changing the opening of a sliding gate (right) until the two intensities are balanced. The use of a red filter allows the comparison to be made at a specific ''color'' (wavelength).

brated tungsten lamp to the incoming unknown source. A filter interposed between the eye and both sources of energy cuts out the shorter wavelengths. This serves a dual purpose: (1) it minimizes the difference between eyes, permitting an easier color match, and (2) it permits an extension of the temperature range beyond the point where the eye could no longer tolerate the amount of energy if viewed directly (Figure 11k).

The instrument is shaped to be held in the hand and up to the eye so that it may be sighted on the target. An adjustable focus permits the operator to focus an image of the source whose temperature is to be determined. The filament of the standard source is placed on the same plane as this image so that the two appear superimposed on one another when viewed through the eyepiece.

A null type of balance is usually used where a rheostat, moving against a calibrated dial, is manually rotated to vary the current through the standard source until it just disappears into the field of the unknown. A slight modification of this principle maintains the standard source constant and varies the amount of interposing absorbing gate opening in the optical path.

The range of the manual optical pyrometer is limited on the low end to a minimum of 1400°F (760°C) since there is insufficient emission of visible light for an accurate comparison below this figure. At 2400°F (1316°C), the image would become too bright to look at directly, but filters are usually interposed to permit readings as high as 6300°F (3500°C).

The use of the human eye as the detector restricts accuracy somewhat because the eye responds to both color and brightness rather than directly to energy, and no two eyes are exactly the same. However, it is possible to detect both a color and a brightness match by adjusting to the ''minimum difference'' between known and unknown.

The manual optical pyrometer is a self-contained unit with its own power supply for operating the current for the known radiant energy source. It can be mounted in place or hand-held by an operator as he or she takes a sighting. Figure 11l illustrates the relationship between distance and target size in using portable infrared pyrometers.

FIG. 11l
Target areas and distances applicable to portable infrared pyrometers. (Courtesy of Wahl Instruments Inc.)

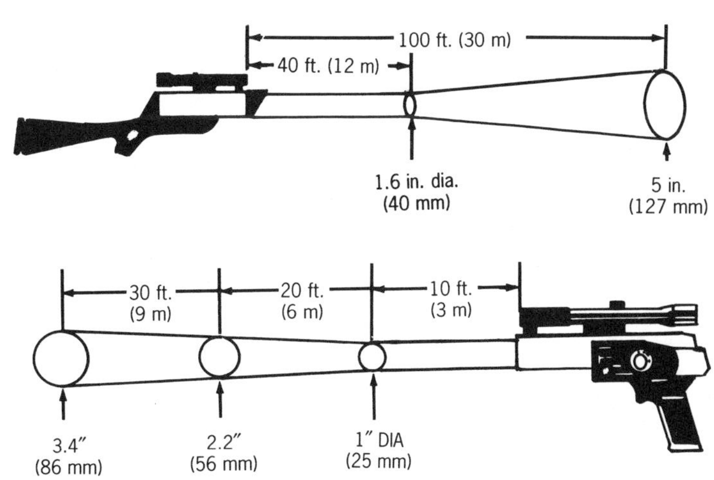

Its advantages include that it is a light, portable, self-contained unit of reasonable accuracy if sighted into a near-blackbody furnace.

Its disadvantages are:

1. It requires the operator to adjust the temperature dial manually. It is not suitable for alarm or control functions.
2. It can only be used at relatively high temperatures where plenty of visible energy is given off.
3. It is subject to emissivity errors inherent to a narrow-band radiation pyrometer.

Automatic Optical and Infrared Pyrometers

The automatic optical pyrometer uses an electrical radiation detector rather than the human eye, and consequently it is not limited to the visible wavelengths of the spectrum. It can reach far into the infrared or the near ultraviolet using either a narrow band, a two-color band, or a wide band selection in accordance with the optical system and detector.

Although there are many adaptations, the automatic optical and/or infrared pyrometer operates essentially by comparing the amount of radiation emitted by the target with that emitted by an internally controlled reference source. The output is proportional to the difference in radiation between the variable source and the fixed reference. The system usually consists of two components: the optical head and the electronic amplifier.

In some models (Figure 11m), the optical head contains a temperature-controlled blackbody source (Figure 11b), the required filters, a detector, a preamplifier, and an optical chopper. The chopper, driven by a synchronous motor, alternately exposes the detector to incoming and cavity radiations at a frequency which might be 60 to 120 cycles per second range.

Another model (Figure 11n) uses the human eye to adjust the focus. Radiant energy passes through the front lens onto a dichroic mirror, which allows visible light to pass through, while infrared radiation is reflected onto the detector. The operator adjusts the eyepiece for the correct focus by just filling the circle reticle plate with the visible image.

The calibrate flag is solenoid-operated from the amplifier and when actuated cuts off the radiation coming through the lens and focuses the calibrate lamp onto the detector. The known radiation from the lamp allows the entire system to be checked.

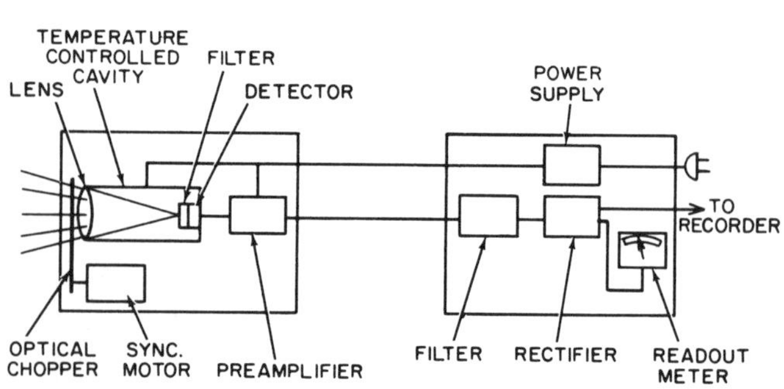

FIG. 11m
Infrared system.

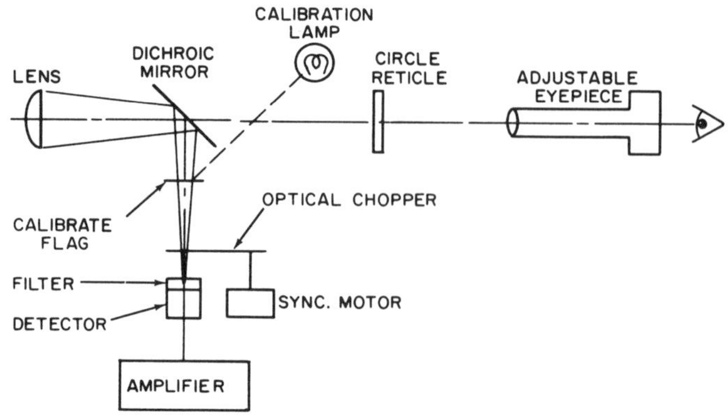

FIG. 11n
System with dichroic mirror.

Automatic optical or infrared pyrometers are closely tied in with radiometers and other optical devices. Telescopic lenses are sometimes employed for their special effects. The instrument may have a wide-angle field of view for large area scanning or it may have a very narrow angle where the target is only .05 in. (1.27 mm) diameter at 8 in. (203 mm) distance.

The entire pyrometer, optics plus amplifier, has been packaged into a single, hand-held, battery-operated device for open fieldwork. It is shaped somewhat like a pistol held in the hand and aimed at the target. Pulling the trigger energizes the standard reference source and the readout indicator (Figure 11l).

A very practical application of the infrared pyrometer in the 8 to 14 micron range is as a hot spot detector. The operator can move quickly from one object to another looking for undesirable internal sources of heat generation.

It is possible to attach a camera to the unit and take a temperature profile as a picture for immediate or future analysis.

TABLE 11o
Uses of Automatic Optical and Infrared Pyrometers

Spectral Response (Microns)	*Main Areas of Application*
0.2–0.3	Jet Engine Rotor Blades, Vacuum Deposition
0.653	Arc Furnaces, Hot Metal Filaments, Refractory Metals
0.6–1.0	Metal Processing Industry
2.0–2.6	Textiles, Plastics, Ceramics, Food Products
3.43 ± 0.14	Thin Film Polymers, Vinyl, Nylon
4.42	Carbon Dioxide Gas
4.8–5.6	Glass, Dry Chemicals, Oxidized Metals
8–14	Background Land, Sea, Clouds, Profiles for Hot Spot Detection

Table 11o shows some typical uses at various spectral responses for this versatile temperature-measuring device.

DETECTORS

The detector receives the radiant energy focused on it by the optical system and generates an electric output signal in response to it. Detectors are grouped into two classes: thermal detectors and photodetectors.

Thermal detectors generate an output because they are heated by the energy they absorb. This category includes thermocouples, thermopiles, pneumatic detectors, the metallic or the thermistor-type bolometers, and the pyroelectric devices. Relative to the photodetectors, these units have a lower sensitivity, and their outputs are less affected by changes in the radiated wavelengths. Most are limited in their speeds of response by their mass; others, like the pyroelectric detectors, are rate-sensitive, meaning that the output responds not to the temperature of the sensor but to the rate at which it is changing. While this speeds up the response, it also requires more electronics and complexity. The use of pyroelectric detectors is limited by their need for optical chopping and their sensitivity to vibration.

Advanced thin-film thermopiles achieve response times in the 10 to 15 millisecond range, while silicon thermopiles also increase the output signal strength. Thermopiles and bolometers respond to radiant energy throughout the whole spectrum and therefore are more suited for detection in total radiation pyrometers. Considering speed, sensitivity, and stability, the best choice for total-radiation applications is the silicon thermopile detector.

Photodetectors are wavelength-sensitive and therefore are better suited to narrow-band pyrometry. The output produced by the photomultiplier tubes and photodetectors is not caused by heat, but by the electrical charges that are released as the radiant energy reaches the detector. The sensitivities of these detectors are in the microseconds. Their main disadvantage is instability at longer wavelengths and when operating at higher temperatures. Consequently, they are used as narrow-band detectors on short-wavelength applications and are frequently provided with cooling.

Photodetectors can be photoconductive, photovoltaic, or photoemissive devices (Figure 11p). The use of photoemissive devices such as photomultiplier tubes has been diminishing. Photoconductive detectors change their resistance as a function of temperature and include the lead selenide and lead sulfide cells. They are sensitive in the 1 to 3 micron range and therefore are either used with filters (Figures 11k, 11m, and 11n) in narrow-band pyrometers or in medium temperature (200 to 800°F, or 93 to 427°C) measurement (Figure 11a) applications as wide-band detectors.

The photovoltaic cell's output is a function of the absorbed radiation. The most widely used cell material is silicone, but germanium and iridium antimonide are also used. The silicon cell matches the short-wavelength thermal emissions of high-temperature objects (0.5 to 1.0 microns) at temperatures from 750 to 7000°F (400 to 3870°C). The germanium cell with its 0.7 to 1.8 micron range is suited for medium temperature measurement and is more stable, more reproducible, and faster than the photoconductive lead sulfide cells. Iridium antimonide has been used to detect the temperature of hot glass surfaces, but due to its poor sensitivity it has been replaced in many applications with pyroelectric thermal detectors in the 8 to 14 micron range. The speed of the photodetectors also permits them to be used for measuring the temperature of small objects moving at high speed, where relative energy output would be small.

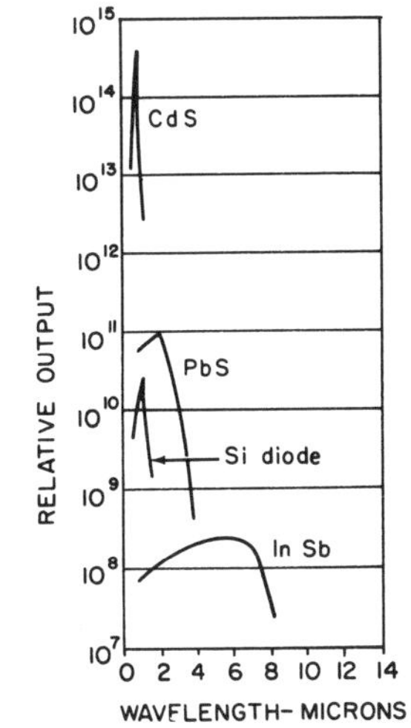

FIG. 11p
Response to energy of some photodetectors.

DESIGN FEATURES

In designing or choosing the radiation pyrometer for a particular application, consideration must be given to the following characteristics:

1. The field of view, or the size-distance relationship.
2. The transmission qualities of the collector system and any windows or filters in the optical path.
3. The band pass and sensitivity of the detector.

Figure 11q shows a typical wide-angle field of view. Note

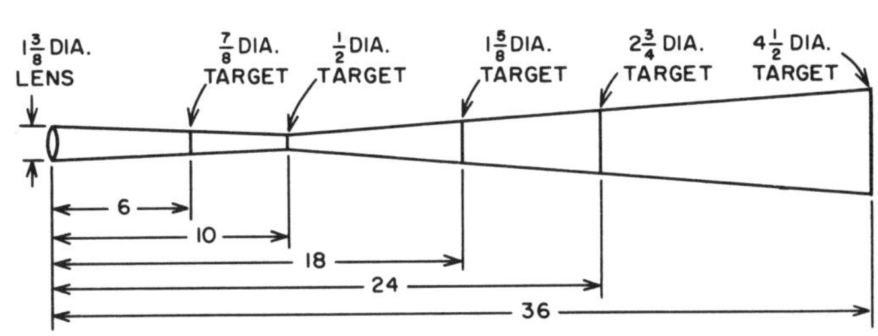

FIG. 11q
Sighting path of the wide-angle total radiation pyrometer.

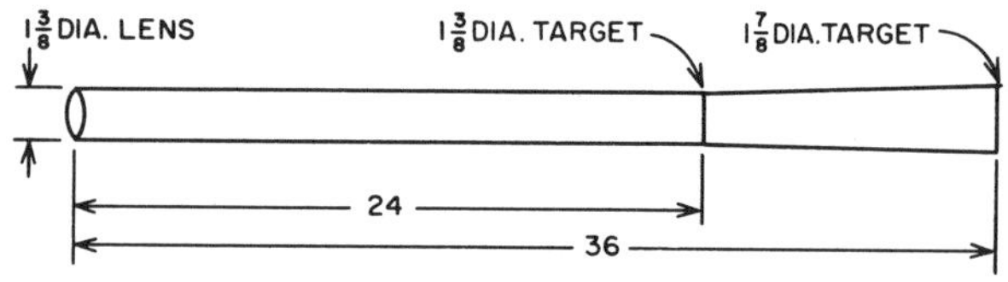

FIG. 11r
Sighting path of the narrow-angle total radiation pyrometer.

how the target size requirement necks down to a minimum at the focal length of a lens in such a system. The narrow-angle field of view, shown in Figure 11r, flares out more slowly. Cross-sectional area in either case can vary from circular to rectangular and even slot-shaped, depending on apertures in the housing design. On some designs telescopic eyepieces can magnify the radiant energy so that much smaller targets at greater distances can be viewed. Targets 1/16 in. (1.6 mm) in diameter are feasible with the proper pyrometer design.

The physical shape of the optical system (lens or curved mirrors) and its mounting within the pyrometer housing control the sighting path, while the material from which it is made determines the optical properties. Glass does not transmit well beyond 2.5 microns and is suitable only for the higher temperatures where plenty of output is available. Other popular optical materials are quartz (fused silica) to 4 microns and crystalline calcium fluoride out to about 10 microns. Lesser used (and more expensive) materials will increase the transmission even more.

Windows and filters in front of or behind the optical system can alter the transmission properties greatly. A plate glass window in front of a calcium fluoride lens, for instance, will very effectively stop the longer wavelengths which would have passed through the lens.

A band pass filter might be purposely placed in front of the detector to cut off unwanted wavelengths.

INSTALLATIONS AND ACCESSORIES

The temperature-measuring requirements for industrial applications involve either the surface temperature of equipment or objects that are out in the open or the temperature inside vessels, pipes, and furnaces. Surface-temperature-sensing pyrometers can be portable (Figure 11l) or permanently installed. When internal temperatures are of interest, one can "look into" the process through windows (Figure 11s). In such installations the pyrometer can be mounted on an adjacent pedestal or supported by a bracket attached to the vessel, as long as the pyrometer is shielded from excessive heat. It is advisable to provide an internal purge to periodically clear the dirt that is likely to accumulate on the window.

When a pyrometer is installed "through the wall" of a furnace a large number of installation-related accessory components are needed. These can easily double the basic cost of the pyrometer itself. In Figure 11t the pyrometer sensor head and its aiming tube, or telescope, are mounted inside the cooling jacket. The coolant flow required is a function of the ambient temperature in the area and varies from 0.2 GPH (0.8 lph) at 150°F (60°C) all the way up to 20 GPH (76 lph) at 600°F (315°C). Proceeding to the left on Figure 11t, the next component is the air purge assembly followed by the safety shutter (or slide gate shutter) which allows the sealing of the furnace when the pyrometer is removed.

Sighting the pyrometer on the target through a tube purged by a clean and nonabsorbing gas minimizes the need for cleaning the lenses and windows. When interference from flames or other sources cannot be filtered out optically, the purged sighting tube, shown on the left of Figure 11t, can be extended almost to the target. In this way the sighting tube allows a more accurate targeting on the point of interest and also makes purging more effective as the air fills the tube. This also reduces the influence of smoke and dust in

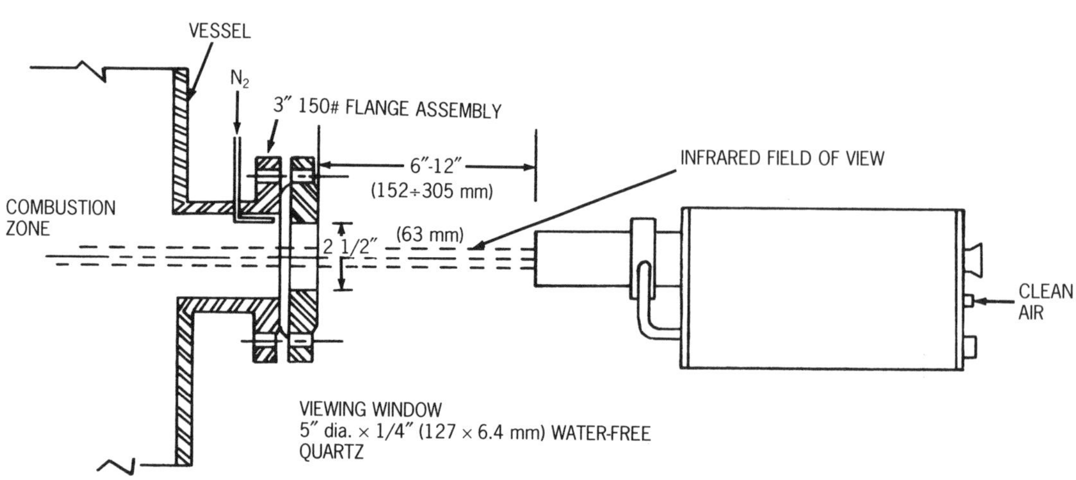

FIG. 11s
Radiation pyrometer installed on a combustion process. (Courtesy of Williamson Corp.)

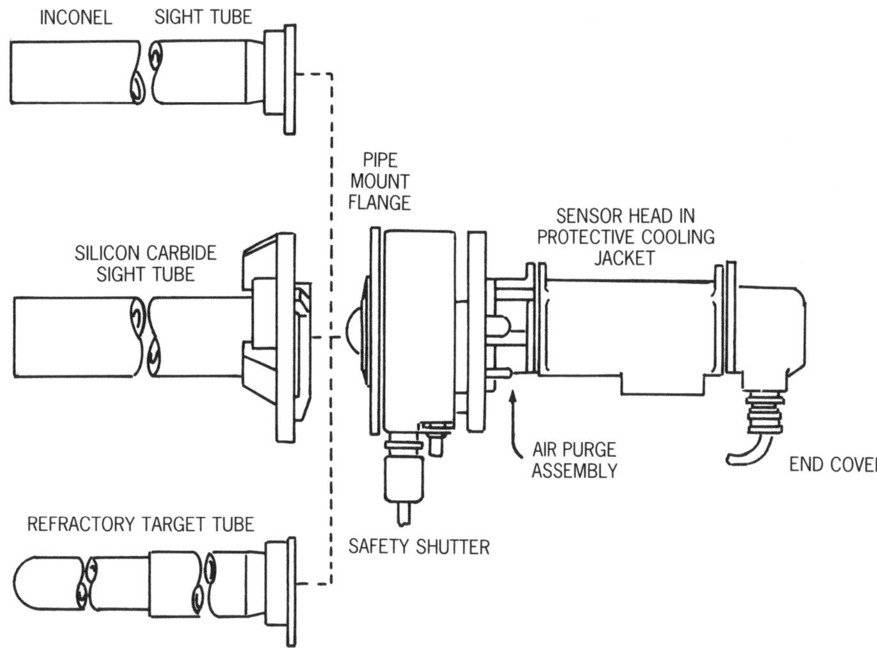

FIG. 11t
Accessories required for installing a pyrometer through the wall of a furnace. (Courtesy of Mikron Instrument Co.)

the atmosphere between the sensing head and the desired target and reduces unwanted background radiation. Sighting tubes are available in different lengths and in constructions of Inconel (for up to 2100°F, or 1105°C), silicon carbide (up to 3000°F, or 1650°C), refractory alumina (up to 2900°F, or 1600°C), or aluminum or stainless steel (below 600°F, or 315°C).

Closed-end "target" tubes (also shown in Figure 11t) are also made of the above listed materials and are used on liquid service or in other processes where the inside temperature of the target tube sufficiently represents the process temperature. In addition to sighting tubes or in place of target tubes, it is possible to form the object of interest into a better or ideal emitter by forming the object into a concave shape or drilling a hole in the target and sighting the pyrometer onto that hole.

ADVANCES AND NEW DEVELOPMENTS

New technologies such as microprocessors, fiber optics, and lasers have also changed radiation pyrometry. One important new development is the availability of the laser reflectometer ("Pyrolaser" is a trademark of Pyrometer Instrument Co.) which can measure the emissivity of the process material at the same location, temperature, and wavelength as used in the pyrometer. This eliminates the potential for emissivity error by allowing its value to be continuously and automatically corrected. This device also helps in overcoming errors caused by radiation from hotter surfaces. This is done by taking a radiant measurement of the hotter refractory surface and compensating for it by calculating its reflectivity as: $R = 1 - E$. As of this writing the laser-based emissivity compensator operates at only one wavelength (0.865 microns) and therefore is effective only if that is the wavelength of interest, but it is likely that with the passage of time it will become available at other wavelengths also.

Other advances include the increase in the number of spectral bands available, the reduction in the required target size (1 mm diameter), and the increase in the distance-to-target ratios, which can exceed 200:1. The "intelligent" pyrometer is more than a sensor—it is a measurement system which routinely provides linear outputs and digital displays and can minimize ambient effects. Intelligent units using two pyrometers are also used to correct for background radiation.

Another area of technological innovation has been in the field of line-scanning pyrometers. These have been developed to monitor the surface temperatures of cement kilns to indicate internal material buildup or refractory deterioration. As the pyrometer scans left to right while the kiln rotates on its horizontal axis, the pyrometer scans its total surface and feeds this information into a small computer for processing. By the same technique, moving sheets of paper, glass, steel, or plastics can be scanned.

SUMMARY

The broad-band pyrometer is used generally in industry for readout and automatic control. It can cover wide temperature ranges and is the least expensive of the three types. Narrow-band and two-color pyrometers are used, where necessary, to minimize the emissivity effects and for special applications where it is desirable to select the particular band pass.

The prospective user of a radiation pyrometer should consider the following points.

1. Target temperature, low, normal and high limits.
2. Minimum target size and distance factors.
3. Target material and emittance.
4. Angle of observation.
5. Is target stationary or moving? If moving, will the speed of response of the pyrometer be fast enough?
6. Atmospheric conditions between target and detector.
7. Ambient temperature.
8. Can pyrometer sight directly on target or must it sight through a sealed auxiliary window such as required for vacuum or pressure?
9. Is scale to be read directly in temperature units or will an arbitrary reading be satisfactory?

ADVANTAGES

1. Does not require physical contact with material whose temperature is being measured.
2. Fast speed of response—can be used on moving targets.
3. Can look at small targets (1⁄16 in. or 1.6 mm in diameter) or measure the average temperature over a wide area.
4. Measures much higher temperatures than thermocouples.

DISADVANTAGES

1. More fragile and costlier than thermocouples, RTDs, or thermistors.
2. Nonlinear scale shape, approximating the 4th power of the temperature.
3. Emissivity of target may cause a low temperature reading if not corrected.
4. Relatively wide temperature span required.

Reference

1. Tenney, A.S., ''Applying Radiation Thermometers to Process and Lab Measurements,'' *InTech,* August 1988.

Bibliography

Adrian, P., ''Technical Advances in Fiberoptic Sensors: Theory and Applications,'' September 1991, pp. 23–45.

Banfield, A.R., ''Evaluating the Performance of Optical and Radiation Pyrometers,'' SPIE Proc., Vol. 234, 1980, p. 62.

Barber, R., ''Review of Infrared Thermometry for the Glass Industry,'' J. Non-Cryst. Solids, Vols. 38 and 39, 1980, p. 903.

Barron, W.R., ''Infrared Thermometry,'' *Measurements and Control,* April 1991.

Bulkley, D., ''Temperature Measurement by Infrared Radiation,'' *Instruments and Control Systems,* September 1970.

de Santoli, L., Ruffino, G., and Santoboni, S., ''Calibration of a Fiberoptics Radiation Thermometer,'' *High Temperature High Pressure,* Vol. 5, 1973.

DeWitt, D.P., et al., ''Theory and Practice of Radiation Thermometry,'' Wiley Interscience, John Wiley, 1988.

''Fiberoptic Sensing of Physical Parameters,'' *Sensors,* October 1988, pp. 21–26.

''Fiberoptic Thermometry for Difficult Surface Temperature Measurements,'' *Sensors,* September 1990, pp. 17–21.

''General Rules for Expression of the Performance of Radiation Thermometers,'' JIS C 1612-1988, Japanese Standards Association, 1988.

Greelish, W.T., ''Non-Contact Temperature Measurement Systems,'' 1976 ISA Conference, Paper #76-807.

Hashemian, H.M., and Peterson, K.M., ''Assuring Accurate Temperature Measurement,'' *InTech,* October 1989.

Hormuth, G.A., ''Ways to Measure Temperature,'' *Control Engineering,* Reprint No. 948, 1971.

Instrumentation and Control Systems Engineering Handbook, TAB Books, 1978.

Intrieri, A.J., ''Optical Fibers Look Around Obstacles to Measure Temperature,'' *Control Engineering,* December 1977.

Katzir, A., Bowman, H.F., Asfour, Y., Zur, A., and Valeri, C.R., ''Infrared Fibers for Radiometer Thermometry in Hypothermia and Hypothermia Treatment,'' *IEEE Transactions on Biomedical Engineering,* Vol. 36, June 1989, pp. 634–637.

Kerin, T.W., and Katz, E.M., ''Temperature Measurement in the 1990s,'' *InTech,* August 1990.

Krigman, A., ''Guide to Selecting Non-contact Temperature Instrumentation,'' *InTech,* June 1983, pp. 23–40.

Leftwich, R.F., ''Infrared Radiometry Applied to Critical Temperature,'' Argonne National Laboratories' 1977 Symposium, Publication No. ANL-78-7.

Merchant, J., ''The Dawn of a New Age in Infrared Thermometry,'' *Measurements and Control,* February 1991.

Ovren, C., Adolfsson, M., and Hok, B., ''Fiber-Optic Systems for Temperature and Vibration Measurements in Industrial Applications,'' Conference on Optical Techniques in Process Control, The Hague, June 14–16, 1983.

Peacock, G.R., ''Radiation Thermometry: Temperature Measurement Without Contact,'' 1991 ISA Conference, Paper #91-0305.

Pepe, R.C., ''Fiberoptic Calibration Considerations,'' NCSL, July 1987, pp. 66-1 to 66-12.

Peterson, W., ''Choosing the Right Temperature Transmitter,'' *InTech,* April 1991.

Saaski, E.W., Hartl, J.C., and Mitchell, G.L., ''A Fiberoptic Sensing System Based on Spectral Modulation,'' ISA. 7/1986/1177-1184.

Sakuma, F., and Hatori, S., ''Establishing a Practical Temperature Standard by Using a Narrow-band Radiation Thermometer with a Silicon Detector,'' TMCSI, 1982, p. 421.

Schoenstein, P.G., ''Infrared Sensors Track Temperature,'' *InTech,* April 1991.

Schooley, J.F., ''State of the Art of Instrumentation for High Temperature Thermometry,'' Argonne National Laboratories' 1977 Symposium, Publication Number ANL-78-7.

Stockdale, R.B., ''New Digital Control Systems Need Better Temperature Sensors,'' *Control Engineering,* November 1991.

Tinsley, F.G., and Adams, B., ''Evolution in the Application of Optical Fiber Thermometry,'' 1991 ISA Conference, Paper #91-0308.

Warren, C., ''Radiation Thermometry,'' *Measurements and Control,* February 1991.

Wickersheim, K., and Sun, M., ''Phosphors and Fiberoptics Remove Doubt from Difficult Temperature Measurements,'' *Research and Development,* Vol. 14, November 1986.

Wohlstein, S., ''Fiberoptics,'' *Measurements and Control,* September 1991.

Wolfe, W.L., *The Infrared Handbook,* U.S. Government Printing Office, 1978.

12 Quartz Crystal Thermometry

T. J. CLAGGETT, R. W. WORRALL (1969, 1982) **B. G. LIPTÁK** (1994)

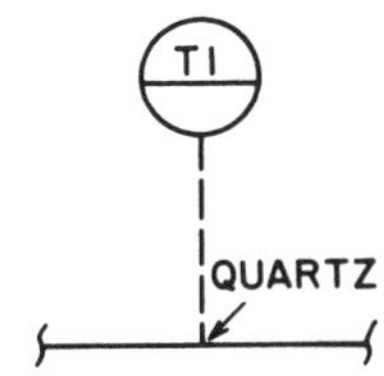

Flow Sheet Symbol

Applications:	Temperature and temperature difference measurements, usually in the laboratory
Temperature Range:	−112 to 482°F (−80 to 250°C)
Inaccuracy:	0.135°F (0.075°C) throughout the full range; 0.072°F (0.04°C) for the range of −58 to 302°F (−50 to 150°C). For temperature difference measurement over an 18°F (10°C) span, the error under laboratory conditions is within 0.018°F (0.01°C)
Cost:	\$5000 for single; \$7500 for differential temperature measurement
Partial List of Suppliers:	Bailey Controls Co.; Hewlett-Packard Co.; Micro Crystal Div. SMH

Temperature measurement using quartz thermometers is based on the change in resonant frequency in response to a temperature change. This temperature sensitivity is on the order of .0005°F (.0003°C) under ideal conditions. Temperature deviations on the order of 10^{-6} °F have been measured in the laboratory.

The effect of temperature on the frequency of quartz crystal controlled oscillators has been known for some time, work on this subject having been published in 1946. Work by various individuals through 1962 resulted in development of the sensitive laboratory device previously mentioned.

A commercial application of the quartz crystal thermometer includes a unique angle of cut, which exhibits a very linear relationship between resonant frequency and temperature. In this case a reference oscillator and a sensor oscillator are employed. The reference oscillator frequency is selected to provide zero beat with the sensor oscillator when the probe is at 32°F (0°C). Sensor oscillators operate to obtain a frequency sensitivity of 500 cycles per second per °F. Dividers are arranged to obtain a sample period of .01 seconds. A resolution of 0.2°F (0.1°C) is obtained with a digital readout (Figure 12a). A digital-analog converter changes the output for use in a strip-chart recorder. A differential measurement can also be made using similar circuitry, with the principal difference being that gating circuits are added to enable the oscillator circuits to be heterodyned against each other.

Probes are provided with the quartz crystal hermetically sealed in a stainless steel cylinder, similar to a well for a thermocouple (Figure 12c) or a resistance thermometer capsule (Figure 12b). This makes the sensor unit larger than either the two sensors mentioned or the thermistor; a ⅜ in. (9.5 mm) outside diameter is typical. However, the probes can be used to pressures of 3000 PSIG (21 MPa) and can stand shocks of 10,000g without changing calibration.

Response time is stated as being 1 second for a step change in a well-stirred water bath. A strong feature is stability, which for short term is in the order of .0001°F (.000056°C). Long-term stability, for periods of a month or

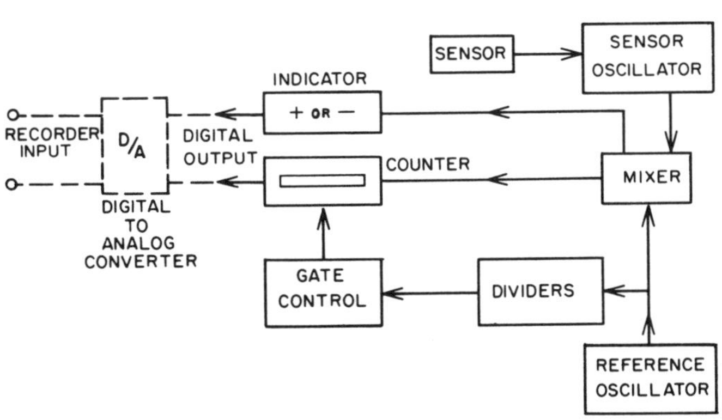

FIG. 12a
Quartz thermometer system block diagram.

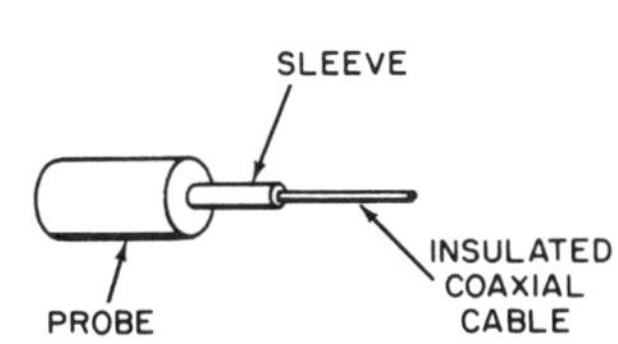

FIG. 12b
Quartz crystal sealed in capsule.

FIG. 12c
Quartz crystal in stainless steel well.

more, is in the order of .02°F (.01°C). These can be degraded by oscillator drift, but, in an environmentally controlled area, this is not much of a problem.

Advantages and disadvantages are summed up as follows:

ADVANTAGES

1. No lead resistance or noise problems because temperature is converted to frequency.
2. Excellent short term stability.
3. Good accuracy.
4. One-second response time.
5. Accurate differential measurements possible.
6. Rugged—can withstand shocks without changing calibration.

DISADVANTAGES

1. Expensive.
2. Accuracy somewhat lower than resistance thermometer or thermistor.
3. Probe size larger than thermocouple or RTD.
4. Best used in laboratory environment.

Bibliography

Hashemian, H.M., and Peterson, K.M., ''Assuring Accurate Temperature Measurement,'' *InTech,* October 1989.

Hormuth, G.A., ''Ways to Measure Temperature,'' *Control Engineering,* Reprint No. 948, 1971.

Kerin, T.W., and Katz, E.M., ''Temperature Measurement in the 1990s,'' *InTech,* August 1990.

Langan, P.E., and Ironside, D.S., ''Calibration Ensures Accurate Temperature Measurement,'' *InTech,* October 1989.

Leewis, W., ''The International Temperature Scale of 1990,'' 1991 ISA Conference, Paper #91-0302.

13 Resistance Temperature Detectors (RTDs)

T. J. CLAGGETT, R. W. WORRALL (1969) **W. A. CLAYTON** (1982)
B. G. LIPTÁK (1993)

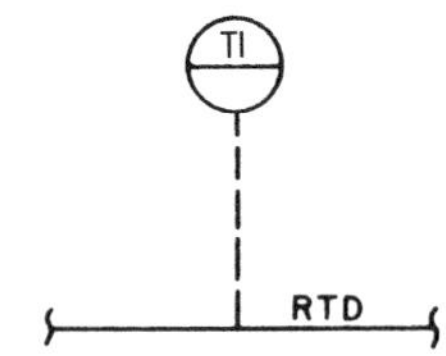

Flow Sheet Symbol

Applications:	Industrial and laboratory thermometry including the definition of the International Temperature Scale between the oxygen and antimony points (−297 to 1167°F, or −183 to 631°C)
Ranges:	−360 to 1560°F (−220 to 850°C) for platinum RTDs and −330 to 600°F (−200 to 320°C) for nickel RTDs
Minimum and Maximum Spans:	10 and 1100°F (6 and 610°C) for platinum, and 20 and 600°F (12 and 340°C) for nickel. For differential temperature transmitters: 10 and 50°F (6 and 28°C) for platinum, and 20 and 100°F (12 and 56°C) for nickel
Stability:	Zero and span drift is usually within 0.1% of span for a 6-month period (and frequently longer)
Standard Resistances:	100 ohms at 32°F (0°C) is most common, but elements are available from 10 to 1000 ohms
Linearity:	Platinum and copper are more linear; nickel and nickel/iron (Balco) are less so. The use of gold and silver RTDs is limited to cryogenic temperatures
Sensitivity:	0.1 to 10 ohms/degree
Inaccuracy:	0.1°F (0.06°C) for platinum sensor alone; 0.15% of span for platinum; 0.25% of span for nickel RTD elements in standard transmitters. Intelligent transmitters give 0.05% of span or 0.18°F (0.1°C), whichever is higher
Costs:	Elements alone start at about $25; a 3- or 4-wire RTD in a stainless steel thermowell costs about $200. An RTD sensor with an integral temperature transmitter costs from $750 to $2000 depending on features
Partial List of Suppliers:	AATech Inc.; Accutech; Action Instrument Co.; Altek Industries Corp.; Analogic Corp.; Anderson Instrument Co.; ARI Industries Inc.; Armtec Industries Inc.; Athena Controls Inc.; Barber Colman Co.; Barnant Co.; Beckman Industrial Corp.; Bristol Babcock; Burling Instruments Inc.; Burns Engineering Inc.; CGS Thermodynamics; Conax Buffalo Corp.; Degussa Corp.; Digitec; Durex International Corp.; Electronic Development Laboratories Inc.; Electronic Instrumentation & Technology Inc.; Enercorp Instruments Ltd.; Enraf-Nonius GmbH; The Foxboro Co.; Gordon Enterprises Inc.; Guildline Instruments Inc.; Hart Scientific Inc.; Heraeus Sensors; Honeywell Inc. Industrial Automation & Control Div.; Hy-Cal Engineering; Instrulab Inc.; JMS Southeast Inc.; Jumo Process Control Inc.; Kube Electronics Inc.; Kyowa Dengyo Corp.; Labfacility Ltd.; Lake Shore Cryotronics Inc.; Leico Industries Inc.; Lewis Electronic Instr. Div. of Colt Ind.; LFE Instruments; Love Controls Corp.; Minco Products Inc.; Moore Industries; Moore Products; Murata Erie North America Inc.; NES Inc.; Newport Electronics Inc.; Omega Engineering; Pyromation Inc.; Pyrometer Instrument Co.; Qualimetrics Inc.; RCD Components Inc.; RdF Corp.; Rochester Instrument Systems; Rosemount Inc.; RTD Co.; Rustrak Instrument Div.; Sarasota Measurements & Controls Inc.; Sealand Sales Ltd.; Senso-Metrics; Solomat Instrumentation; S-Products Inc.; Statham Transducer Div. of Schlumberger Industries; Temco Products Co.; Temperature Measurement Systems; Temp-Pro Inc.; Testoterm Inc.; Thermalogic Corp.; Thermo-Couple Products Co.; Thermo Electric Co.; Transmation; Tudor Technology Inc.; United Electric Controls Co.; Validyne Engineering Corp.; Wahl Instruments Inc.; Weed Instrument Co.; West Coast Research Corp.; Westcon; West Instrument, Mark IV; Wilkerson Instrument Co.; Yellow Springs Instrument Co.; Yokogawa Corp. of America

In 1821 Sir Humphry Davy announced that the resistivity of metals depends on their temperature. Resistance thermometry is based upon the increasing electrical resistance of conductors with increasing temperature (Figure 13a). The application of this property using platinum was first described by Sir William Siemens at the Bakerian Lecture of

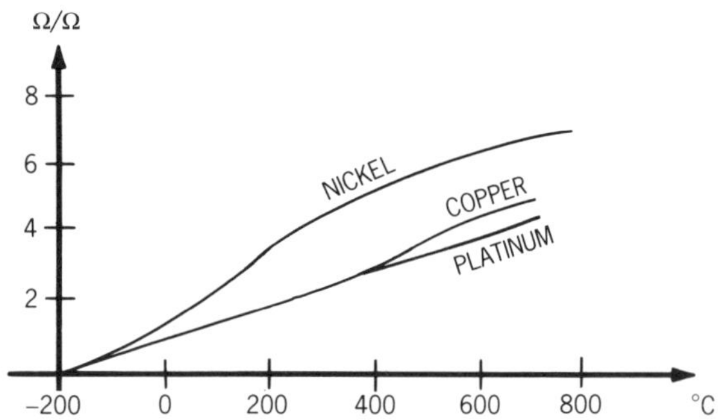

FIG. 13a
The temperature coefficient of platinum is 0.00392 ohms/ohms-°C at 0°C and 0.00293 at 800°C. Its average value is 0.00385 between 0 and 100°C. Therefore (as shown) a platinum element of 1 ohm resistance at −200°C will increase its resistance to approximately 4.0 ohms as the temperature rises to 800°C.

1871 before the Royal Society in Great Britain. The necessary temperature limitations and methods of construction were established by Callendar, Griffiths, Holborn, and Wein between 1885 and 1900. Platinum resistance thermometers are now the international standard for temperature measurements between the triple point of hydrogen at 24.86R (13.81K) and the freezing point of antimony at 1167.35°F (630.75°C). The laboratory application of platinum resistance thermometers recognizes the unsurpassed stability and repeatability of this noble metal sensor. Platinum resistance thermometers of more rugged industrial construction retain this advantage over other conductors used in industrial resistance thermometers.

The conductors used for resistance thermometry include platinum, nickel of various purities, 70% nickel/30% iron (Balco™), and copper, listed in order of decreasing temperature range capability (Table 13b). No other candidates are generally available in commercial thermometers. These conductors are all available as fine wire for sensor winding. Platinum is also available as a deposited film sensor, and nickel and Balco are available in foil-type sensors. All share in varying degrees the characteristics of repeatability, high temperature coefficient, long-term stability, and linearity over a useful temperature range. While no sensor material surpasses platinum in overall performance, each has at least one characteristic that may encourage its selection.

The application of platinum as a temperature standard (SPRT), including the careful application of manual bridge resistance measurements, differs considerably from industrial practice.[1] The temperature standard is 25.5 ohms at the ice point to stay within the range of practical Mueller bridges (Figure 13h) while providing a nominal 0.1 ohm/°C sensitivity. Standard platinum thermometers are constructed in a manner to be almost totally strain-free, using very lightly supported wire of larger size (and cost) than typical in an industrial thermometer (RTD) (Figure 13c). Temperature coefficients near the theoretical maximum for pure platinum and maximum thermal stability are obtained at the expense of fragility and large size. The industrial platinum RTD offers ruggedness at a negligible loss in temperature coefficient compared to the standard SPRT, while retaining the best stability and repeatability available in industrial thermometers over a wide temperature range.

INDUSTRIAL RTD CONSTRUCTION REQUIREMENTS

All resistance thermometers require the following considerations in their manufacture. Wire-wound sensors must be supported on mandrels closely matching the wire in thermal expansion to minimize strain effects. Additional assembly materials, such as cements, should not introduce additional strain in the operating temperature range. The final assembly must be in a stable, annealed condition, trimmed to the required resistance tolerance. Only high-purity materials and clean assembly methods should be used to avoid sources of contamination that might degrade the sensor. All internal connections should be welded, and connecting leads should be chosen for the required temperature capability and avoidance of thermoelectric junctions. To realize the ruggedness of fully supported elements in the total sensor assembly, all internal connections should be anchored and isolated from

TABLE 13b
RTD Material Properties at 0°C and Available Temperature Ranges

Metal	*Temperature Coefficient of Resistance (Ohms/Ohm°C)*	*Electrical Resistance Per Circular Mil Foot (Ohms)*	*Commercially Available Temperature Ranges °F (°C)*
Platinum	0.00392	59.00	−420 to 1650 (−251 to 899)
Nickel	0.00672	36.00	−320 to 600 (−196 to 316)
Copper	0.0038	9.26	−320 to 250 (−196 to 120)
Tungsten	0.0045	30.00	—
Gold	0.0040	13.00	—
Silver	0.0041	8.80	—

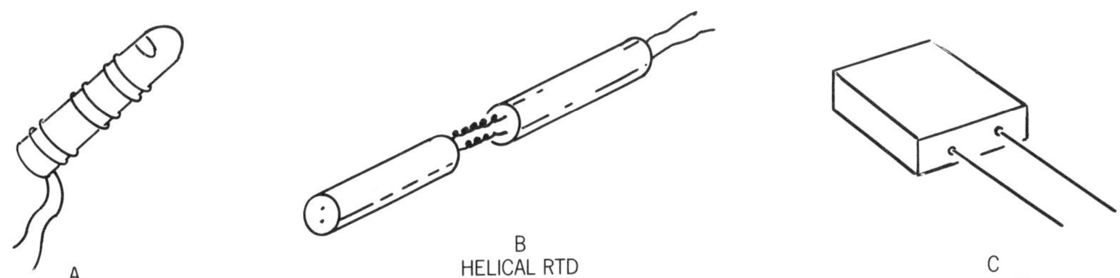

FIG. 13c
RTD elements can be made by winding platinum wire on a glass or ceramic bobbin and sealing with molten glass (A), by threading a platinum helix through a ceramic cylinder (B), or by depositing a platinum film on a ceramic substrate. (A) is limited by the strain induced at higher temperatures, (B) is not suited for extreme vibration, and (C), while small, fast, and inexpensive, is also less stable than the others.

effects of thermal and mechanical strains, including shock and vibration. The same requirements apply when deposited film or foil-type resistance elements are used (Figure 13c).

For equivalent performance in their respective temperature ranges, base metal RTDs cost the same as 100 ohm platinum RTDs. Construction requirements and materials cost are similar. Base metal RTDs have a materials cost advantage at higher resistance values compared to wire-wound platinum sensing elements. Thin film platinum elements erase this advantage.

Platinum RTDs

In the case of platinum RTDs, the fully supported rugged construction using ''reference grade'' wire has a temperature coefficient (alpha) over the interval 32 to 212°F (0 to 100°C) of between .00387 and .003915 ohms/ohm-°C, depending on the manufacturer. Compared to .003927 ohms/ohm-°C on a SPRT, the reduction in sensitivity is insignificant. For best accuracy, the user should be aware of or specify the actual temperature coefficient. One common value available from most manufacturers is .003902 ohms/ohm-°C. This is the result obtained for windings on pure alumina mandrels.

Standard platinum industrial RTD curves based on slightly doped platinum wire have been adopted by several countries. These curves are all substantially identical with an alpha of .00385 ohms/ohm-°C. This result is reproducible by manufacturers everywhere, and the so-called international grade platinum curve is the most widely used curve, even in the United States. An exception is temperatures below −320°F (−196°C) (LN_2), where only reference grade wire RTDs are well characterized.

Platinum RTDs using thick or thin films are also available to the same curve as international grade wire-wound sensors. Performance is often equivalent to wire-wound sensors except maximum temperature may be reduced. Wire-wound platinum RTDs are most common at 100 ohms ice point resistance, with 200 and 500 ohms available at additional cost. Using thin films, ice point resistances of 100 and 1000 ohms are available at the same cost with slightly lower alpha specified at 1000 ohms.

Base-Metal RTDs

Second in usage to platinum is high-purity nickel, which offers the highest temperature coefficient, second-highest temperature range, and lower assembled cost than wire-wound platinum at high resistance values. The most common are 120 and 500 ohm resistances, with 1000 ohms available. Nickel is nonlinear, with an increasing temperature coefficient at increasing temperature (Figure 13a). Nickel is also highly strain-sensitive and requires great care by the manufacturer to obtain good interchangeability. The temperature coefficient of nickel is highly influenced by both purity and state of anneal. In addition, lower purity nickel such as 99% nickel (no longer available) or ballast nickel has been used and provides a somewhat lower temperature coefficient. The maximum temperature of nickel sensors should not be much more than 500°F (260°C). There is no internationally recognized standard curve for nickel sensors, although there are national standards, and several manufacturers in the United States can provide sensors to a common curve characterized by an alpha temperature coefficient between 32 and 212°F (0 and 100°C) of .00672 ohms/ohm-°C.

Third in usage is the 70% nickel/30% iron alloy tradenamed Balco. The sole basis for its use is a very high specific resistance which makes possible very high resistance windings without increasing size. Ice point resistances are commonly 2000 or 10,000 ohms. It has the second-highest temperature coefficient, third-highest temperature capability, and, like pure nickel, is nonlinear with an upward bending R versus T curve. There is no recognized standard curve for Balco sensors.

Last in usage is pure copper, which is generally available only at 10 or 100 ohm ice point resistance values due to low specific resistance of the winding wire. Copper's temperature coefficient is almost the same as platinum and it is very linear above the ice point. Copper in bifilar windings (Figure

TABLE 13d
Comparison Chart of Various Temperature Sensors

Evaluation Criteria	*Platinum Rtd 100Ω wire wound and thin film*	*Platinum Rtd 1000Ω thin film*	*Nickel Rtd 1000Ω wire wound*	*Balco Rtd 2000Ω wire wound*	*Thermistor*	*Thermocouple*	*Semi-Conductor Devices*
Cost-OEM Quantity	HIGH	LOW *	MEDIUM	MEDIUM	LOW *	LOW *	LOW *
Temperature Range	WIDE −400°F to +1200°F (−240°C to +649°C) *	WIDE −320°F to +1000°F (−196°C to +538°C) *	MEDIUM −350°F to +600°F (−212°C to +316°C)	SHORT −100°F to +400°F (−73°C to +204°C)	SHORT to MEDIUM −100°F to +500°F (−73°C to +260°C)	VERY WIDE −450°F to +4200°F (−268°C to +2316°C) **	SHORT −57°F to +257°F (−49°C to +125°C)
Interchangeability	EXCELLENT **	EXCELLENT **	FAIR	FAIR	POOR to FAIR	GOOD *	FAIR
Long-Term Stability	GOOD *	GOOD *	FAIR	FAIR	POOR	POOR to FAIR	GOOD to FAIR
Accuracy	HIGH *	HIGH *	MEDIUM	LOW	MEDIUM	MEDIUM	MEDIUM
Repeatability	EXCELLENT **	EXCELLENT **	GOOD *	FAIR	FAIR to GOOD	POOR to FAIR	GOOD *
Sensitivity (output)	MEDIUM	HIGH *	HIGH *	VERY HIGH **	VERY HIGH **	LOW	HIGH *
Response	MEDIUM	MEDIUM to FAST *	MEDIUM	MEDIUM	MEDIUM to FAST *	MEDIUM to FAST *	MEDIUM to FAST *
Linearity	GOOD *	GOOD *	FAIR	FAIR	POOR	FAIR	GOOD *
Self-Heating	VERY LOW to LOW *	MEDIUM	MEDIUM	MEDIUM	HIGH	N/A	VERY LOW to LOW *
Point (end) Sensitive	FAIR	GOOD *	POOR	POOR	GOOD *	EXCELLENT **	GOOD *
Lead Effect	MEDIUM	LOW *	LOW *	LOW *	VERY LOW **	HIGH	LOW *
Physical Size/Packaging	MEDIUM to SMALL	SMALL to LARGE *	LARGE	LARGE	SMALL to MEDIUM *	SMALL to LARGE *	SMALL to MEDIUM *

** Best Rating
* Good Rating

13c) is used in electrical machinery due to very low inductive or capacitive reactance, but platinum can also be used. Some traditional applications have also exploited the linearity of copper sensors in narrow range temperature difference measurements where two sensors are connected on opposed arms of a bridge. There is no internationally recognized standard curve for copper, although some national standards exist.

Relative performance of industrial temperature sensors of all types is shown in Table 13d. RTDs, especially platinum RTDs, are clearly superior for temperature measurement requirements below 1000°F (540°C).

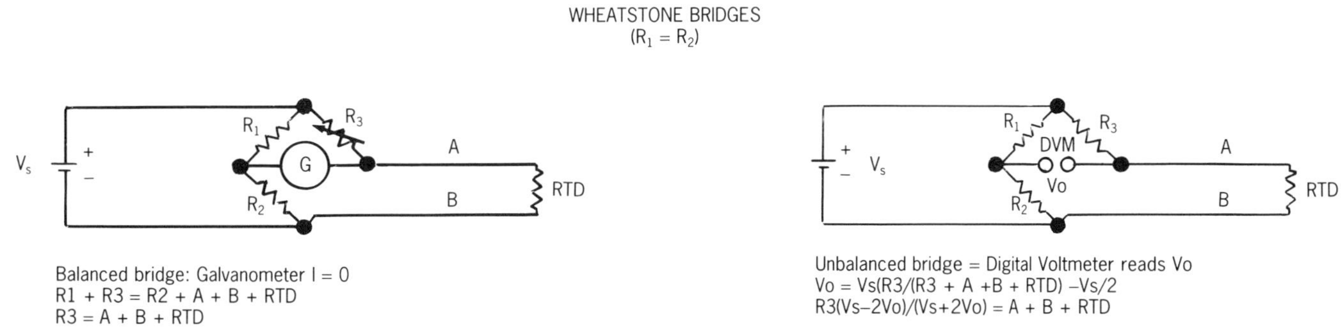

FIG. 13e
Null-balance-bridge-type two-wire RTD installations showing a galvanometric (G) readout in a balanced condition (left) *and a digital voltmeter readout in an unbalanced condition* (right) *where the DVM reads V_O volts.*

MEASURING THE RTD RESISTANCE

By measuring the resistance of the RTD element one can determine the process temperature if the change in total resistance measured is affected by nothing but the process temperature. In actual installations the RTD element is connected by lead wires to the readout or transmitting instrument. If a null-balance bridge is used the circuit can be as shown in Figure 13e. Here the temperature coefficients of resistors R_1, R_2, and R_3 are nearly zero and the value of R_3 is adjusted until the current flow of G or the voltage read by a digital voltmeter (DVM) is zero. Under these conditions the value of R_3 equals the unknown resistance of the measuring leg (A + B + RTD).

Assuming that the resistance of the RTD (at 0°C) is 100 ohms and assuming that the element is platinum, the resistance of the 100 ohm RTD element will change by 0.385 ohms for each degree Centigrade change in temperature. If 500 ft (152 m) of 20 gauge copper lead wire was used to connect the RTD to the bridge, this adds 10 ohms to the total resistance (5 ohms per leg). At null balance R_3 = A + B + RTD = 10 ohms + RTD. With a temperature coefficient of 0.385/°C these extra 10 ohms will contribute a measurement error (as an increase in process temperature) of 10/0.385 = 26°C.

The previous example illustrates the relatively large lead-wire error in a two-wire RTD installation, and for this reason such installations are not used if accurate temperature measurement is desired and the length of the lead wires is more than a few inches. When the transmitter is mounted directly on the thermowell (Figure 13f) the lead-wire length is so small that the resulting error is not significant. Yet even in these configurations the better suppliers will provide a three-wire RTD to minimize the lead-wire error.

Three-Wire RTD

As shown in Figure 13g the lead wire C acts as a sense lead and is part of both halves of the bridge and therefore cancels out at balance. The lead wires A and B are in different halves of the bridge and therefore at null balance R_3 = B − A + RTD. Therefore, now the lead-wire error is no longer the

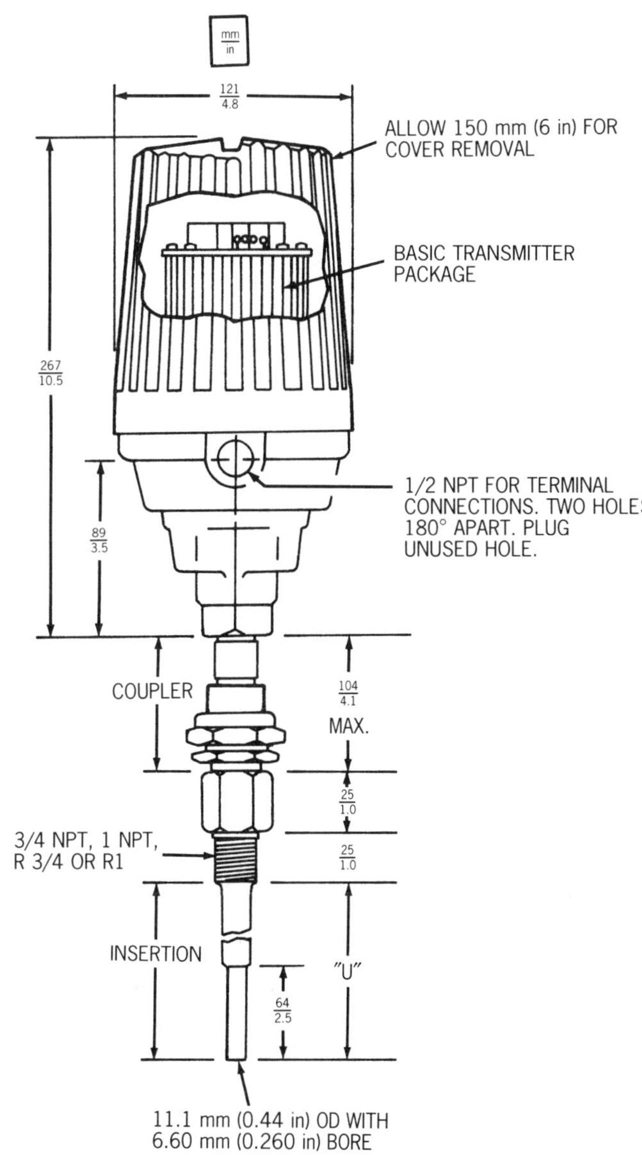

FIG. 13f
RTD transmitter mounted directly on thermowell. (Courtesy of The Foxboro Co.)

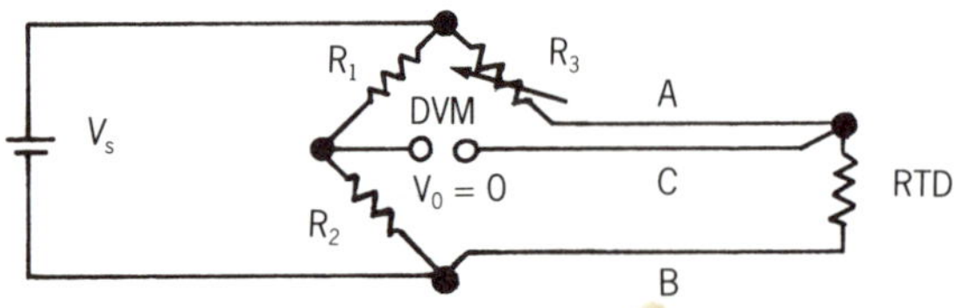

When bridge is balanced: R1 + R3 + A + C = R2 + B + RTD + C

If R1 = R2 this becomes: R3 = RTD + B − A = RTD

FIG. 13g
In three-wire null-balance bridge the lead-wire effect is reduced to the difference between the resistances of the two lead wires (B − A).

total lead resistance (A + B), but only the difference between their resistances (B − A). This is a major improvement in reducing the lead-wire error and is sufficient for the needs of most industrial applications where the lead-wire lengths are short. However, it is not a complete solution because wire resistances are guaranteed only within a 10% tolerance; therefore, if A and B are identical wires of identical lengths their resistances can still differ within the 10% tolerance. So if nominally they both are 5 ohms, in reality one could be 4.5 and the other 5.5 ohms. If this were the case, the difference of 1 ohm would still introduce an error. With a 100 ohm platinum RTD that error would correspond to 1/0.385 = 2.6°C.

If the purpose of the temperature measurement is to calculate the exothermic heat release of a batch reactor, this error might still be too much. In that application the temperature rise through the reactor jacket is about 5°F and the span usually selected for the differential temperature transmitter is 10°F (5.6°C). In order to identify the end point accurately, the total heat release must be determined to within 0.5% maximum error. Because the total heat release is calculated by multiplying coolant flow with its temperature rise, the flowmeter itself will contribute 0.25% in error and therefore one must measure the temperature rise within 0.25%. An error of 0.25% over an *actual* measurement of 5°F is 0.0125°F (0.007°C). This is such a small error limit that even three-wire RTD transmitters may not meet it (their usual error limit is about 10 times higher). For this reason, in laboratory situations or for other high-precision measurements one might consider the use of four-wire systems, which completely eliminate the lead-wire effect.

Four-Wire RTDs

Four-wire RTDs can be connected either to a null-balance bridge or to a constant current source. Both will be described here. Figure 13h illustrates a four-wire null-balance bridge. It operates by switching a triple-pole double-throw switch and making alternate null-balance measurements in the two configurations. In one configuration lead A is measured together with the RTD resistance, while in the other configuration it is lead B, so they cancel out *completely* and the actual value of the RTD resistance is determined as $(R_{3a} + R_{3b})/2$. Microprocessors and advanced electronics make it feasible to provide this level of sophistication, but complexity still costs money, so these designs are relatively expensive; in addition, they are still limited by contact resistance considerations. Even the best (gold-plated) switching contacts contribute some contact resistance, and the difference between these resistances does introduce some miniscule errors whenever one uses a switching configuration to make a resistance measurement.

Another way to eliminate the lead-wire error is to use a constant current source (CCS) in a four-wire RTD configuration. These miniaturized CCS packages are available at relatively low costs and provide an accurately constant current flow of about 2 milliamperes or less to avoid self-

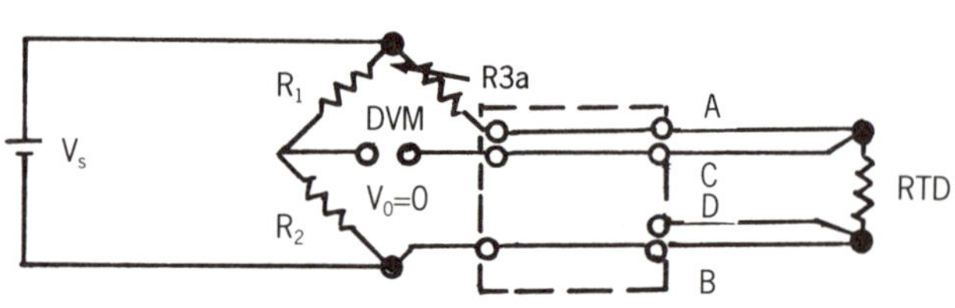

SWITCH IN POSITION "A"

R1+R3a+A+C = R2+B+RTD+C
When R1=R2:
R3a + A = B + RTD

FIG. 13h
If the leads of a four-wire null-balance bridge are switched as shown and the resulting two equations are added up, the lead-wire effects are eliminated and the resistance of RTD = $(R_{3a} + R_{3b})/2$.

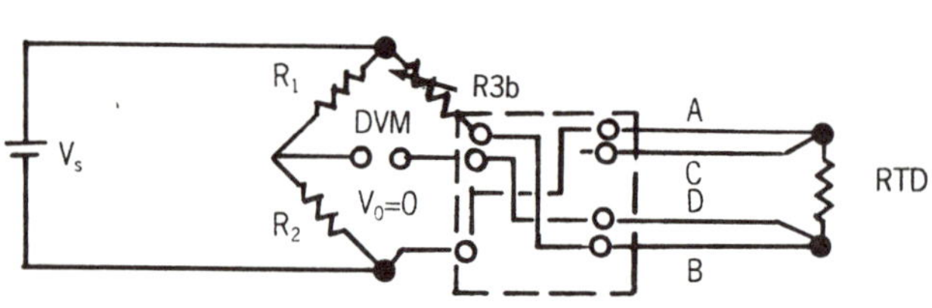

SWITCH IN POSITION "B"

R1+R3b+B+D = R2+A+ RTD+D
When R1=R2
R3b + B = A + RTD

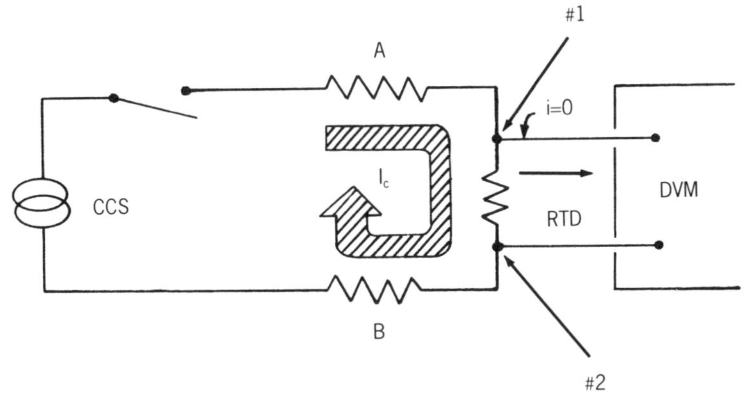

FIG. 13i
Offset compensated four-wire RTD measurement using constant current source (CCS) and digital voltmeter (DVM) as readout. (Courtesy of Hewlett-Packard)

heating errors. As shown in Figure 13i, in this configuration the bridge itself is replaced by a digital voltmeter which measures the resistance of only the RTD and is insensitive to lead-wire effects as there is no current flow through the connecting wires. The source lead resistances (A and B) contribute no error because the voltage drop is not measured along them.

For the maximum in precision, it would be prudent to make sure that the current flow (I_c) through the RTD is constant and that the DVM does draw any current ($i = 0$), and also to cancel out the thermocouple junction voltages at points #1 and #2. This is necessary because as the two wires (platinum RTD and copper lead) at #1 and #2 form thermocouple junctions, the millivoltages they generate will also be registered by the DVM. This effect is eliminated by offset compensation. The offset voltage generated by the unintended thermocouple junctions is measured by the DVM when the CCS circuit is opened and, therefore, $I_c = 0$. The ''smart'' RTD readout memorizes the voltage sensed when no current is flowing and corrects the total reading by that amount when the CCS is connected and I_c is about 2 milliamperes.

In general, two-wire RTDs are used in HVAC-type secondary applications, three-wire RTDs are used in the processing industries, and four-wire RTDs are used in high-precision services or in the laboratory.

SENSOR CONSTRUCTION

Figure 13c illustrated the RTD sensor elements used. The most popular industrial designs are the fully encapsulated RTDs. In these units a 0.001 in (0.025 mm) diameter or smaller platinum wire is wound into a coil and is inserted in the multiple bores of a ceramic tube or is wound directly on the outside of a ceramic tube. The ceramic material is usually 99.7% pure aluminum oxide, and the winding is completely embedded and fused onto or into the tube. The RTD thermometers are in direct competition with thermocouples and therefore are available with the same features.

Figure 13j illustrates an industrial RTD sensor provided with a protective thermowell, while Figure 13k shows two designs of surface-temperature-sensing RTDs. The packaging of the RTD sensor depends on the application. For example, the measurement of temperature inside glass-lined

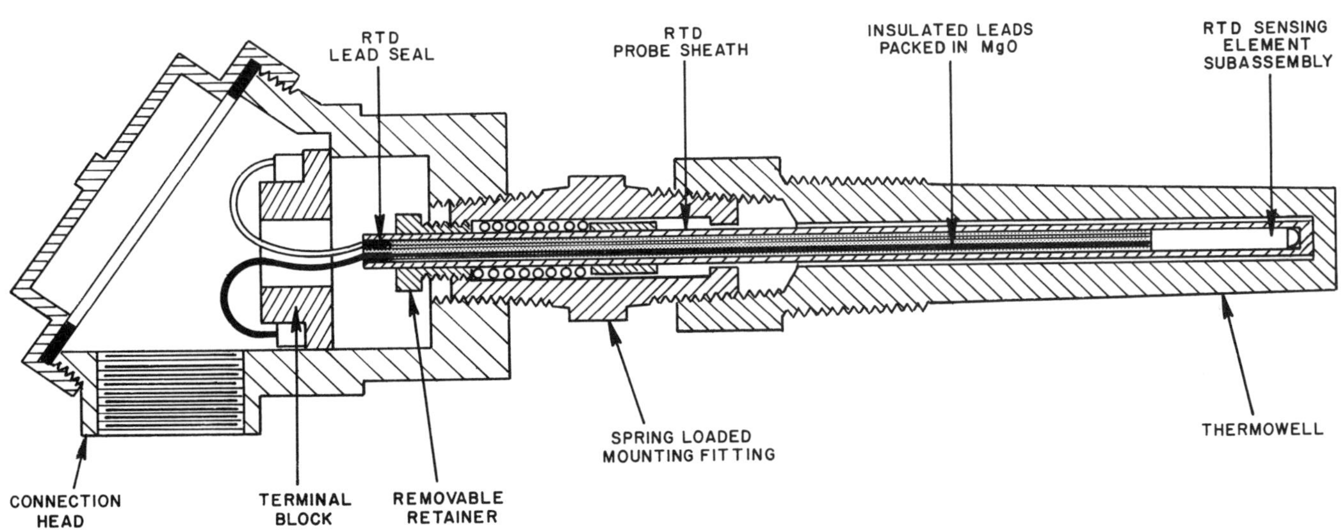

FIG. 13j
Industrial RTD/thermowell assembly.

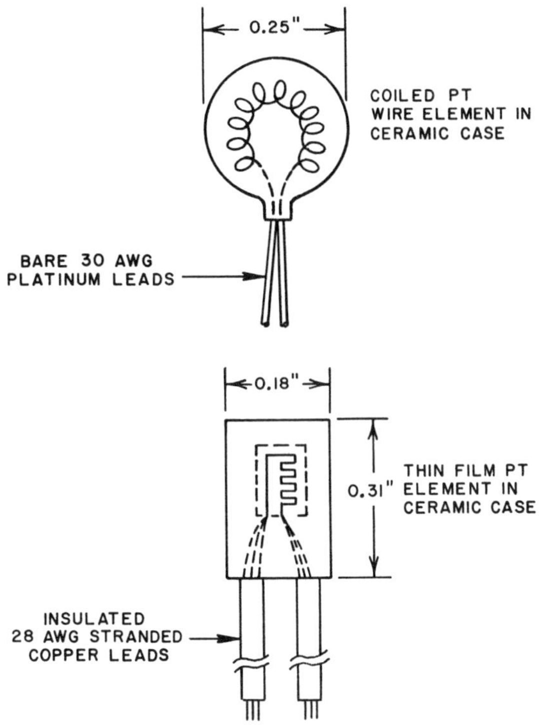

FIG. 13k
RTD surface temperature sensors.

chemical reactors requires special configurations. Figure 13l shows an RTD element that can be installed either through the bottom discharge valve of the reactor or at the tip of the reactor baffle. The 63.2% response time to a step change is about 7.5 seconds; the 95% response time is about 28 seconds.

One limitation of RTDs (relative to thermocouples) is size. RTDs tend to be relatively bulky, because in order to obtain the required resistance (usually 100 ohms) the length of the sensor wire must be relatively long, frequently several feet. This limitation has been overcome by the film-type designs (Figure 13c), which are suited for miniaturization. The 63.2% time constants of different RTDs (without thermowells) are given in Figure 13m. One might approximate the time constant of the final assembly by doubling the time constants for each layer of material added in the process of installation.

STANDARD AND INTELLIGENT TRANSMITTERS

The typical performance capabilities of different RTD transmitters are summarized in Table 13n. In general, platinum sensors provide better accuracy and less span shift than do nickel elements. Also, while intelligent transmitters give better performance than do standard ones, they too are limited to a minimum error of 0.18°F (0.1°C). Table 13n also shows that the smart transmitters that have digital outputs provide slightly better performance.

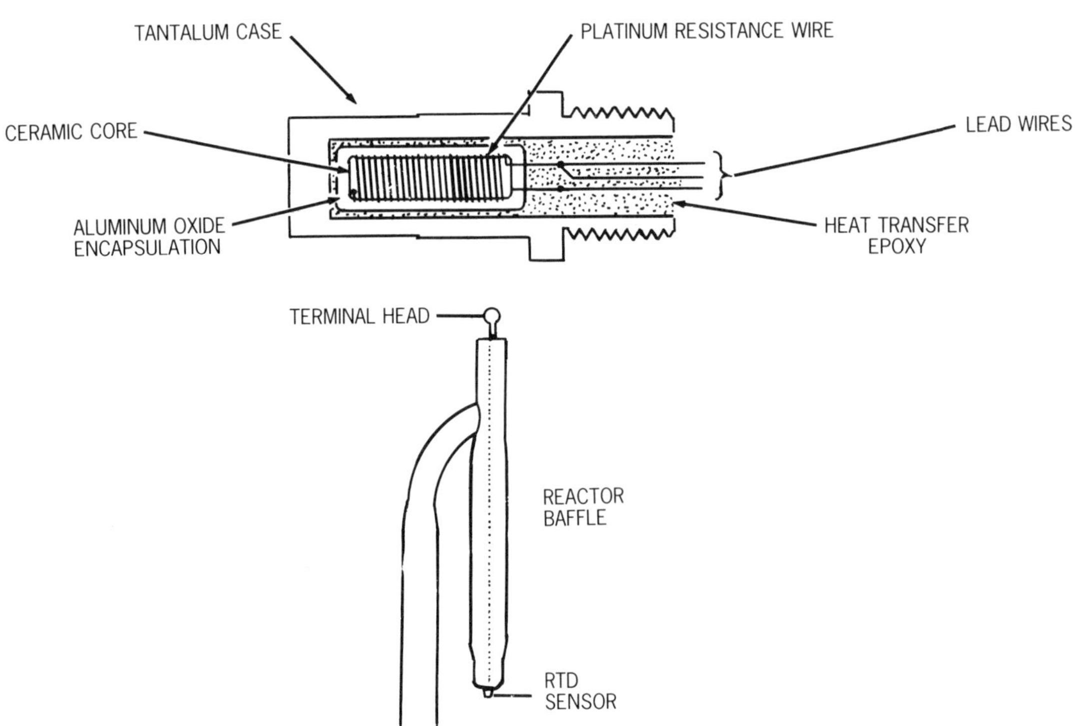

FIG. 13l
RTD sensor for glass-lined reactors. (Courtesy of Pfaudler)

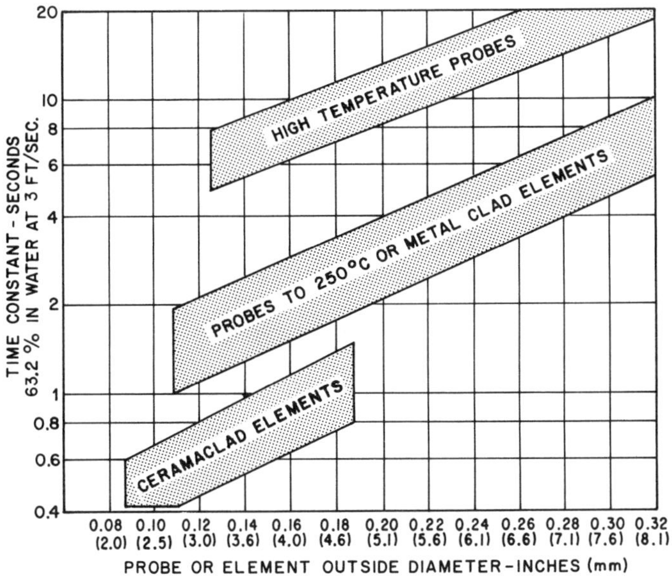

FIG. 13m
Response time of typical resistance temperature sensor.

In addition to having improved performance, intelligent transmitters are capable of working with any one of eight types of thermocouples or two types of RTD elements. This increases their flexibility and reduces the need for spare parts. The intelligent transmitters are also provided with continuous self-diagnostics and with automatic three-point self-calibration which is performed every 5 seconds and does not interrupt the analog or digital output of the unit. The intelligent RTD transmitter can also be furnished with dual RTD elements that can be used to measure temperature differentials, averages, or high/low sensors, or used as redundant backup elements. Another convenient feature of smart transmitters is their remote reconfiguration capability, which can change their zero, span, or many other features without requiring rewiring.

ADVANTAGES AND LIMITATIONS

RTDs are among the most accurate, reproducible, stable, and sensitive thermal elements available. Some of the precision platinum RTDs can measure within a few thousandths of a degree, and this precision is the reason such instruments are used to define parts of the International Temperature Scale (ITS-90). Other advantages include their relatively good sensitivity (0.1 to 10 ohms/°F) and their ability to use conventional copper lead wire (instead of more expensive thermocouple wire). An advantage of copper RTDs is that since both the element material and the lead-wire material are the same, the thermocouple effect is minimized at their junction. Yet another advantage of RTDs is the convenience of using a single bridge to measure the temperature difference between two RTDs.

The disadvantages of RTDs include their higher cost, more fragile construction, and larger size, relative to thermocouples. Because of their size, their thermal response time is also relatively slow (Figure 13m). RTDs are also subject to self-heating. The size of this error rises with RTD size and resistance and is reduced by heat transfer and by minimizing (about 2 milliamperes in CCS) or eliminating (null-balance) the current flow through the RTD. Errors can be introduced

TABLE 13n
*Performance Capabilities of Standard and "Smart" RTD Transmitters**

	STANDARD		*SMART*	
Performance Criteria	*Platinum Element*	*Nickel Element*	*Digital Output*	*Analog Output (4–20 mA DC)*
Inaccuracy	±0.15% or ±0.15°F (0.08°C)	±0.25%	±0.035% or ±0.18°F (0.1°C)	±0.05% or ±0.18°F (0.1°C)
Repeatability	±0.05%	±0.05%	±0.015% or ±0.18°F (0.1°C)	±0.025% or ±0.18°F (0.1°C)
Zero Shift/6 mo.	±0.1%	±0.2%	±0.06% R or 0.18°F (0.1°C)	±0.1% R or 0.18°F (0.1°C)
Span Shift/6 mo.	±0.1%	±0.4%		
Supply Voltage Variation	±0.2% or 0.02°F (0.01°C)		—	(0.005%)/Volt
Ambient Effect (100°F, or 55°C)	±0.75%		Included above	±0.1%

*When two values are given the error is the higher of the two. When % is given it refers to % of span or % of calibrated span, except if %R is shown, which means % of actual reading.

if the RTD insulation resistance is affected by moisture being sealed in the sheath or due to contact between element and sheath. Some RTDs are more vibration-sensitive than others. RTDs also depend for their precision on stable (insensitive to temperature changes) and constant resistances and power supplies in the associated bridges.

In terms of installation, RTDs require the same precautions as thermocouples. The best installation practice is to place all electronics directly on top of the thermowell (Figure 13f) and thereby eliminate lead-wire and noise effects. If for some reason this cannot be done, the lead wires should be twisted and shielded; also, the wires should not be stressed, strained, or made to go through steep gradients. The extension wire should be low resistance (large diameter), and the readout instrument should be guarded.

Reference

1. Riddle, J.L., Furukawa, G.T., and Plumb, H.H., "Platinum Resistance Thermometry," NBS Monograph 126, 1973.

Bibliography

American Petroleum Institute Manual Number API RP 550.

Anderson, R.J., "The High Temperature Stability of Platinum Resistance Thermometers," in *Temperature, Its Measurement and Control in Science and Industry,* Vol. 4, Part 2, Research Triangle Park, North Carolina: ISA.

Baker, H.D., Ryder, E.A., and Baker, M.A., *Temperature Measurement in Engineering,* Vol. II, New York: John Wiley and Sons, Inc., 1961.

Ball, K., "Thermocouples and RTD's," *InTech,* August 1986.

Bediones, D., et al., "Criteria for the Selection of Thermocouples Versus RTDs in Industrial Applications," 1991 ISA Conference, Paper #91-0300.

Carlson, D.R., "Transmitting a Remote Temperature," *Control,* April 1989.

Corruccini, R.J., "Interpolation of Platinum Resistance Thermometers, 10° to 273.15°K," in *Temperature, Its Measurement and Control in Science and Industry,* Vol. 3, Part I, New York: Van Nostrand Reinhold, 1962.

Evans, J.P., et al., "A Study of Stability of High Temperature Platinum Resistance Thermometers," New York: Van Nostrand Reinhold, 1962.

Hormuth, G.A., "Ways to Measure Temperature," *Control Engineering,* Reprint No. 948, 1971.

Instrumentation and Control Systems Engineering Handbook, TAB Books, 1978.

The International Practical Temperature Scale of 1968, Amended Edition of 1975, Metrologia, 12, July 17, 1976.

Kerin, T.W., "Temperature Measurement in the 1990s," *InTech,* August 1990.

Leewis, W., "The International Temperature Scale of 1990," 1991 ISA Conference, Paper #91-0302.

Meyers, C.H., "Coiled Filament Resistance Thermometers," *NBS Journal of Research,* Vol. 9, 1932.

Petersen, W., "Choosing the Right Temperature Transmitter," *InTech,* April 1991.

"Platinum Resistance Temperature Sensors," Bulletin 9612, Rosemount Engineering.

"Resistance Thermometers, Thermistors," *Measurements and Control,* April 1991.

Schooley, J.F., "State of the Art of Instrumentation for High Temperature Thermometry," Argonne National Laboratories' 1977 Symposium, Publication No. ANL-73-7.

Standard DIN 43760, Fundamental values of measuring resistors for resistance thermometers, German Standards Committee (DNA).

Tavener, J.P., "Platinum Resistance Temperature Detectors," *Measurements and Control,* April 1974.

Trietley, H.L., "Avoiding Error Sources in Platinum Resistance Temperature Measurement," *InTech,* February 1982.

14 Temperature Switches and Thermostats

B. G. LIPTÁK (1982, 1993)

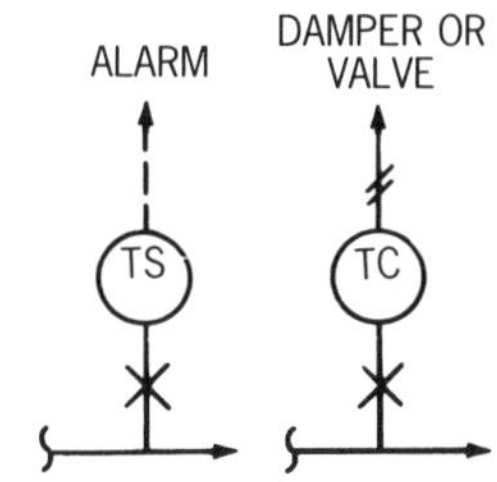

Flow Sheet Symbol

Types:	A. Temperature switches B. Thermostats
Range:	A. From 0 to 75°F (−18 to 24°C) up to 50 to 650°F (10 to 343°C) spans are standard B. Room thermostat setpoints are adjustable from 45 to 85°F (7 to 29°C) for heating and from 55 to 105°F (13 to 40°C) for cooling services
Inaccuracy:	A. 0.5 to 2% of span B. Traditional thermostats are usually uncalibrated and their performance is not guaranteed. The microprocessor-based programmable thermostats are repeatable within 1°F (0.6°C)
Costs:	Bimetallic spring elements cost only a few dollars; conventional room thermostats cost about $100; programmable microprocessor-based energy-saver thermostats cost $200 to $400; general purpose temperature switches are $150; industrial-quality, indicating, explosion-proof temperature switches with thermowells cost $300 to $500
Partial List of Suppliers:	Allen-Bradley Co. (A); Ametek Inc. (A); Aschroft-Heisse Div. of Dresser (A); Automatic Switch Co. (A); Bailey Controls Co. (A); Barber Colman Co. (B); Barksdale Controls Div. of Imo Industries (A); Burling Instruments Inc. (A); Custom Control Sensors Inc. (A); Dresser Industry Instrument Div. (A); Dwyeer Instruments Inc. Mercoid Div. (A); Elcon Instruments Inc. (A); Emerson Electric Co. (B); Fenwal Inc. (A,B); Fluid Components Inc. (A); Honeywell Inc. (A,B); ITT Industrial Controls (A,B); Johnson Controls Co. (A,B); Jumo Process Control Inc. (A); Kobold Instruments Inc.; Leeds & Northrup (A); MTL Inc. (A); Omega Engineering (A); Powers Process Control (A,B); PSG Industries Inc. (A); Robertshaw Controls (A,B); Rochester Instrument System (A); SOR Inc. (A); Square D Co. (A,B); Stafea Control System (A,B). Therm-O-Disc Inc. (A); Thermo Electric Ltd. (A); 3D Instruments (A); United Electric Controls Co. (A); Weiss Instruments Inc. (A); Weksler Instruments Corp. (A)

Temperature switches are on-off devices while thermostats are narrow (frequently fixed) proportional band controllers. Their common characteristic is that measurement, setpoint, and control functions are all combined into a single instrument.

TEMPERATURE SWITCHES

Temperature switches are used to energize and de-energize electric circuits as a function of the relationship between the process, temperature, and a predetermined setpoint. They are available with accuracies up to ±½% of span. The sensing elements are mostly the elastic types including filled and bimetallic elements. The electric switching assemblies are either snap-acting mechanical microswitches or mercury switches. The latter contain no mechanical moving parts and must be mounted on a vibration-free level surface.

Figure 14a illustrates some of the terminology used in connection with temperature switches. The temperature range within which the actuation point can be set is referred to as *adjustable range.* The switch may actuate at its setpoint on rising (*high*) or falling (*low*) temperature. The setpoint is the temperature that actuates the switch to open or close an electric circuit. The setpoint accuracy defines the band within which repetitive actuations will occur. Differential or dead band is the difference between setpoint and reactuation

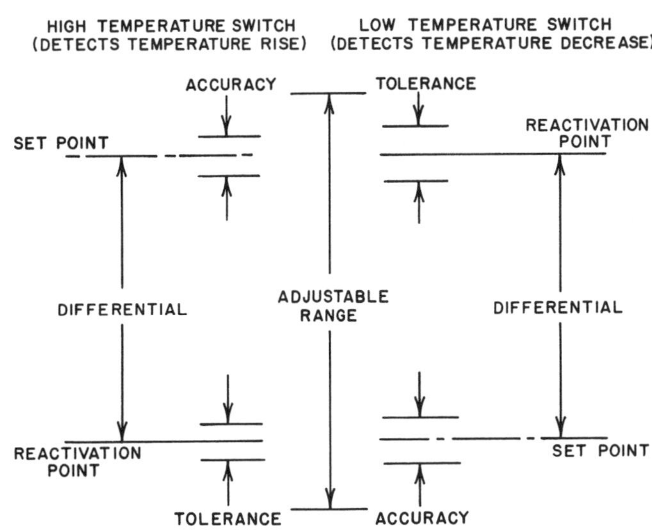

FIG. 14a
Temperature switch terminology.

point. For example, if a high temperature switch is set to actuate (close) at 100°F (38°C) on rising temperature, it will close at that point. When the temperature drops, it will not open again until the temperature has fallen to 95°F (35°C). In this case, the differential is 5°F (2.8°C). Tolerance is the repeatable accuracy of the reactuation point.

Temperature switch elements should be selected with service life and maximum operating temperature in mind. Most elastic elements will have a service life of close to a million cycles if the cycle time is not less than five seconds. Selection of the adjustable range for a specific installation should consider both the setpoint actuation accuracy and the life factor. For greatest accuracy, the setpoint should fall in the upper half of the range, but for longest service life it should be in the lower half. The usually acceptable compromise is to locate the setpoint in the middle third of the range.

It is desirable to have an external calibrated knob provided on the temperature switch for setpoint adjustment. Uncalibrated or internal setpoint adjustments are generally undesirable on industrial installations. The "fixed differential" temperature switches are furnished with a single adjustment for setpoint. These units are factory-set with close differentials amounting to ½ to 1% of span. On double-adjustment-type designs, both set and reactuation points can be independently adjusted. The maximum differential in such designs is the range of the switch, while the minimum varies between 2 and 8% of span. Temperature switches with dual control are also available. Here, two independent switches mounted in the same housing are responding to the same process temperature in opening or closing two independent circuits.

The electrical rating of temperature switches at a 115 volt operating level varies from 0.3 to 10 amperes on AC or DC circuits. Generally, the dual control and the fixed differential switches have lower ratings, and the double-adjustment-type units have higher ratings. The available circuit arrangements are very flexible. Some of the standard arrangements include single-pole–single-throw (SPST), single-pole–double-throw (SPDT), and double-pole–double-throw (DPDT) designs, but units are available with up to four poles.

There are basically three standard case designs: general purpose (NEMA 1), weather-resistant (NEMA 2 and 3), and explosion-proof (NEMA 7) cases.

In some designs, the control mechanism is an integral part of the explosion-proof case, while in others it can be removed in the field for maintenance. As noted earlier, the setpoint should be externally adjustable so that the explosion-proof case does not need to be opened in order to change the setpoint. For added convenience of the operator, some switches also provide continuous process temperature indication. Another optional feature is push-button reset, which must be manually operated before the circuit will be restored to its original state after an automatic operation. For example, the operation of a low temperature switch with "manual reset" would be such that if it is to start a heating device when the low temperature setting is reached, it will not stop that device when normal temperature is restored, but will keep it in operation until reset manually.

THERMOSTATS

There is little difference between a two-position thermostat and a temperature switch. The temperature-sensing elements in the thermostats shown in Figure 14b are bimetallic springs. In the design shown on the left, the gap between contacts is almost zero and the movement of the bimetallic element directly opens and closes the load contact. In the design shown in the center a positive gap is provided, serving to reduce wear on control equipment. The gap is produced by toggle action and is adjustable by adjusting the spring tension. In slow processes with long cycle periods, such as in domestic thermostats, "anticipation" is provided through the use of a negative gap (right of Figure 14b). In this thermostat, when the heater in a room is turned on a small internal heater within the thermostat is also actuated. This causes the bimetallic spring to turn off the room heater before the room temperature reaches the temperature inside the thermostat.

The next level of sophistication is to replace the completely on-off action (with or without adjustable gap) of the thermostat with a small amount of throttling. In such throttling thermostats the room temperature has to move away from setpoint by a degree or two before the heating (or cooling) valve (damper) is fully opened. One simple way of achieving such throttling action is by letting the bimetallic spring move a flapper in front of a nozzle (Figure 14c), thereby causing the pneumatic output pressure of the thermostat to change with temperature.

Thermostats are distinguished from other controllers in that they provide proportional action only and that their proportional band (gain) is *narrow and fixed.* A conventional room thermostat might have a setpoint range of 55 to 85°F (13 to 29°C) and a *fixed sensitivity* (gain) of 2.5 PSI/°F (0.3 bar/°C). Assuming that this thermostat operates a control valve having a 9 to 13 PSIG (0.6 to 0.9 bar) spring range, we can convert the fixed sensitivity of 2.5 PSI/°F into a percent proportional band value that is better understood by instrument engineers.

Sensitivity (or gain) is stated as the ratio between a change in measurement and a change in the corresponding

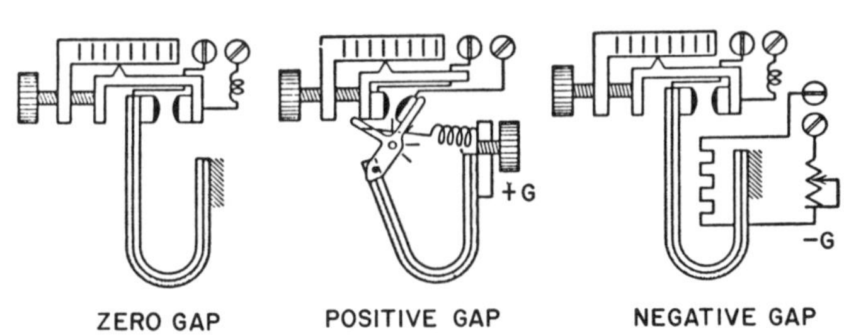

FIG. 14b
Design variations of electric two-position thermostats.

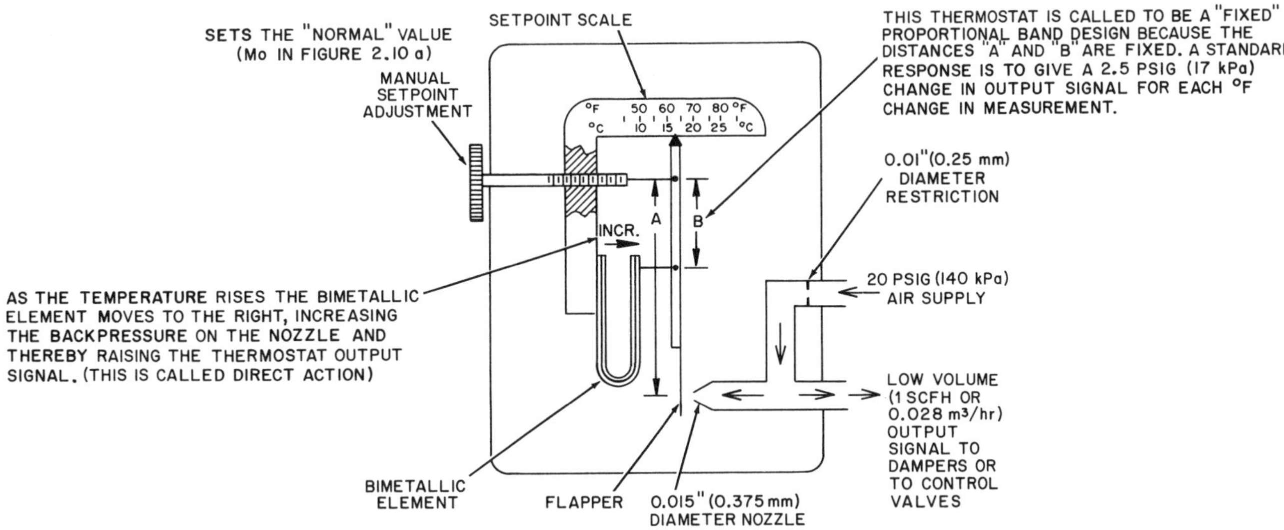

FIG. 14c
Direct-acting bimetallic thermostat with manual setpoint and fixed proportional band. The output air flow is limited because there are no relays, so this is called a "low-volume" design.

controller output. A 1°F (0.56°C) change in measurement is 3.3% of the measurement range of 30°F (55–85°F, or 13–29°C), while the 2.5 PSI change in controller output is 62% of the controller output range of 4 PSI (9–13 PSIG). If a 3.3% change in input results in a 62% change in output, that is a gain of 62/3.3 = 19. Converting this typical thermostat gain to proportional band, we get:

$$P = \frac{100}{G} = \frac{100}{19} = 5.3\% \qquad 14(1)$$

This controller, having a fixed proportional band of 5%, is very sensitive, has a small "offset," and, as its gain is fixed, cannot be tuned.

Let us examine the consequences of these characteristics separately. A sensitive controller is suited for the control of very slow, large-capacity processors. Temperature control in the HVAC industry usually fits that description and therefore thermostats with fixed, high gains can give acceptable results, if space temperature can change only very slowly. On the other hand, such thermostats cannot control spaces with fast dynamics (short time constants) and will cycle or lose control when applied to such service.

The "offset" inherent in all proportional controllers is not a serious drawback because the resulting error is small, due to the narrowness of the proportional band. Using our earlier example and assuming that the thermostat output is set at 11 PSIG (0.76 bar) when the error is zero (thermostat on setpoint), we can calculate the maximum error due to offset. When there is no error, the control valve is 50% open (output is 11 PSIG, or 0.76 bar). As the load changes, an error must be allowed to develop in order for the valve opening to change. Assuming that the valve has been correctly sized and is large enough to handle all expected loads, the maximum error due to offset will be that deviation from setpoint which is required to move the valve from half to full opening. This 2 PSIG (0.14 bar) change in thermostat output will occur when the deviation from setpoint is 0.8°F (0.4°C).

Therefore, this is the size of the permanent offset error for a gain of 19:1. There is an inverse linear relationship between gain and offset error, such that if the gain is cut in half, the offset error will be doubled.

From the above discussion it might be concluded that the conventional room thermostat is a good selection for the HVAC-type applications and that more expensive instruments, such as PID controllers, would not necessarily improve the overall performance. While that is true, it should also be noted that conventional thermostats are usually uncalibrated devices and their manufacturers usually do not guarantee their accuracy. This is a limitation, because it is possible to have some thermostats with 5 to 10°F (3 to 6°C) error in their measurement. The HVAC industry is now benefiting from a new generation of thermostats, with guaranteed measurement error limits of 1°F (0.6°C) or less.

Design Features

Application engineers can choose from a fairly large variety of design features when specifying thermostats:

Thermostats can be electric or pneumatic.
They can be indicating or blind.
They can have direct or reverse action.
Some thermostats will change from direct- to reverse-acting in response to a change in air supply pressure.
The measuring element can be bimetallic, filled, or electronic.
The thermostat setpoint can be local or remote.
Local setpoints can be external, concealed, or accessible only through the use of a key.
If the setpoint of a pneumatic thermostat is adjusted remotely by an air signal, each 1 PSIG (0.07 bar) change in

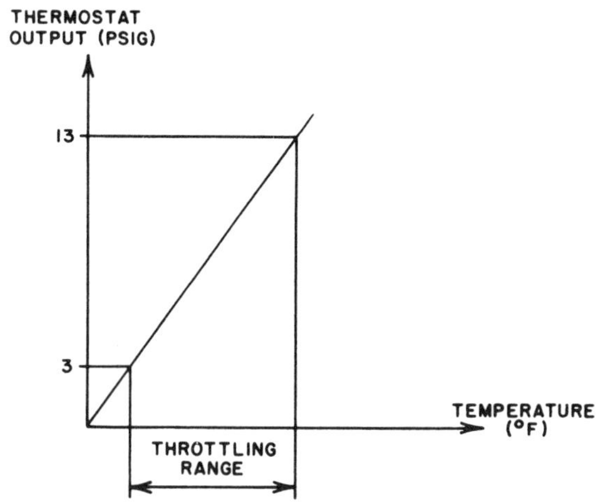

FIG. 14d
Definitions of throttling range.

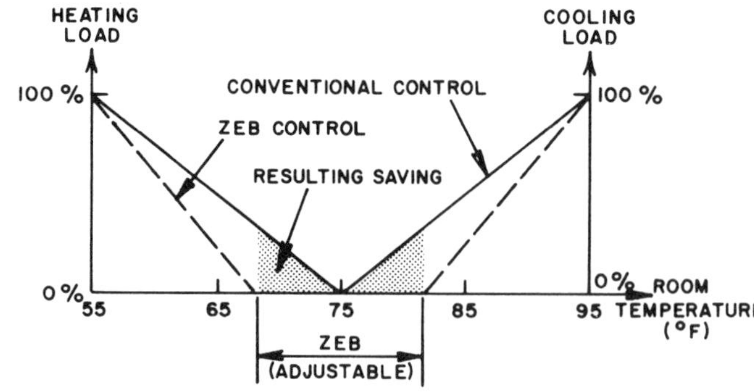

FIG. 14e
Zero Energy Band (ZEB) control.

setpoint pressure will move the setpoint by an adjustable preset amount. The range of this adjustment is usually from 0.15 to 1.4°F (0.1 to 0.8°C) per 1 PSI (0.07 bar).

If the setpoint is to change as a function of the time of day, a timer can automatically operate a solenoid and thereby switch the setpoint signal.

Some of the more recently developed and more advanced thermostat features include:

Adjustable gains, or proportional bands. Another term used to describe the sensitivity of thermostats is *throttling range.* As shown in Figure 14d, it refers to the amount of temperature change that is required to change the thermostat output from 3 to 13 PSIG (0.2 to 0.9 bar). The throttling range is usually adjustable from 2 to 10°F (1 to 5°C).

Dual setpoints. These thermostats will switch their settings in response to a change in the air supply pressure. Both setpoints can be manually adjustable, with the day setting made by external thumbwheel and the night setting concealed internally.

''Limited control range.'' These thermostats allow the occupant of an office to move the setpoint to any value desired, but will disregard any setting that exceeds the limit value. For example, in heating applications, the limit could be 74°F (23°C). In this case the space temperature will be limited to a maximum of 74°F, regardless of the setting by the occupant. Similar limit values can be set for cooling.

Zero Energy Band Control

A recent addition to the available thermostat choices is the Zero Energy Band (ZEB) design. The idea behind ZEB control is to conserve energy by not using any when the room is comfortable. As illustrated by Figure 14e, the conventional thermostat wastes energy by continuing to use energy when the area's temperature is already comfortable. The ''comfort gap'' or ''ZEB'' is adjustable and can be varied to match the nature of the particular space involved.

ZEB control can be accomplished in one of two ways. The single setpoint and single output approach is illustrated on the left side of Figure 14f. Here the cooling valve fails closed and is shown to have an 8 to 11 PSIG (0.55 to 0.76 bar) spring range, while the heating valve is selected to fail open and has a 2 to 5 PSIG (0.14 to 0.34 bar) range. Therefore, between 5 and 8 PSIG (0.34 and 0.55 bar) both valves are closed, and no pay energy is expended while the thermostat output is within this range. The throttling range is usually adjustable from 5 to 25°F (3 to 13°C). Thus, if the ZEB is 30% of the throttling range, it can be varied from a gap size of 1.5°F (.85°C) to 7.5°F (4.2°C) by changing the throttling range (or gain).

While the split-range approach is a little less expensive than the dual setpoint scheme (shown on the right of Figure 14f), it is also less flexible and more restrictive. The two basic limitations of the split-range approach are:

1. The gap width can only be adjusted by also changing the thermostat gain; maximum gap width is limited by the minimum gain setting of the unit.
2. In this design, the heating valve must fail open which is undesirable from an energy conservation point of view.

These limitations are removed when a dual setpoint, dual output thermostat is used. Here both valves can fail closed and the band width is independently adjustable from the thermostat gain. The gains of the heating and cooling thermostats are also independently adjustable. In Figure 14f, the heating thermostat is reverse-acting and the cooling thermostat is direct-acting.

The most recent advances in thermostat technology are the microprocessor-based units. These are programmable devices with memory capability. They can be monitored and reset by central computers using pairs of telephone wires as the communication link. Microprocessor-based units can be provided with continuously recharged backup batteries and with accurate (transistor-type) room temperature sensors. They can also operate without a host computer (in the ''stand alone'' mode). In this case the user manually pro-

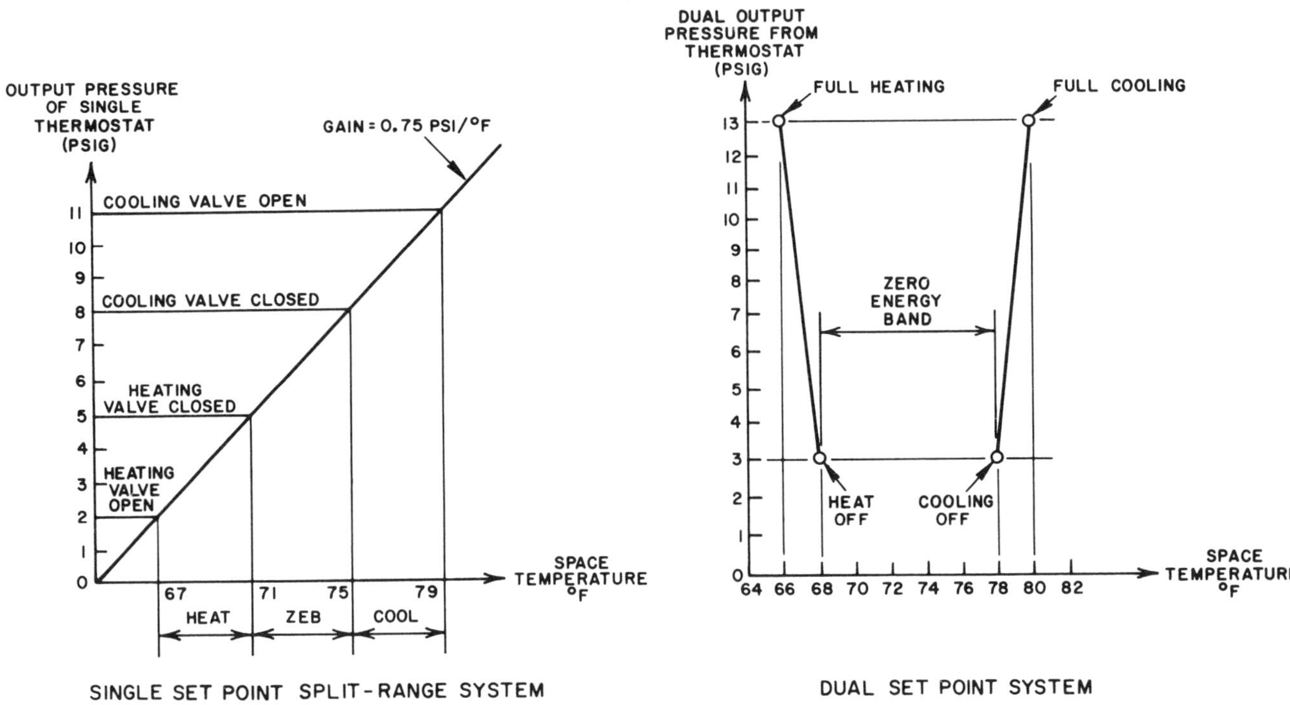

FIG. 14f
Implementation of ZEB control schemes.

grams the thermostat to maintain various room temperatures as a function of the time of day and other considerations.

Bibliography

ASHRAE Handbook and Product Directory, 1980 Systems, Chapter 34 on "Automatic Control."

"Filled-Systems Thermometers," *Measurements and Control,* September 1991.

Hashemian, H.M., et al., "Assuring Accurate Temperature Measurement," *InTech,* October 1989.

Jutila, J.M., "Temperature Instrumentation," *Instrumentation Technology,* February 1980.

Krigman, A., "Guide to Selecting Temperature Switches," *InTech,* June 1984.

Lipták, B.G., "Reducing the Operating Costs of Buildings by the Use of Computers," ASHRAE Transactions, Vol. 83, Part I, 1977.

Plumb, H.H., "Temperature, Its Measure and Control in Science and Industry," 5th Symposium on Temperature, held in 1971 by NBS, API, and ISA.

Roots, W.K., *Fundamentals of Temperature Control,* New York: Academic Press, 1969.

Spethmann, D.H., "Importance of Control in Energy Conservation," ASHRAE Journal, February 1975.

Eckman, D.P., *Automatic Process Controls,* New York: John Wiley & Sons, 1958.

"Gas Appliance Thermostats," AGA (American Gas Association) Standard: ANSI: Z21.23-1975.

"Method of Testing for Capacity Rating of Thermostatic Refrigerant Expansion Valves," ASHRAE Standard: 17-75 (ANSI: B60.1-1975).

"Residential Controls—Integrally Mounted Thermostats for Electric Heaters," NEMA (National Electrical Manufacturers' Association) Standard: DC13-1973.

"Residential Controls—Low Voltage Room Thermostats," NEMA Standard: DC3-1972.

"Standard for Fire-Detection Thermostats," UL (Underwriters' Laboratories) Standard: UL-521 (ANSI: Z220.1-1971).

"Standard Method of Test for Flexibility of Thermostat Metals," ASTM (American Society of Testing and Materials) Standard: B106-68 (1972).

"Standard Method of Test for Thermal Deflection Rate of Spiral and Helical Coils of Thermostat Metal," ASTM Standard: B389-65 (1975).

"Thermostatic Refrigerant Expansion Valves," ARI (Air-Conditioning and Refrigeration Institute) Standard: 750-76 (1976).

When, S., "Liquid-in-glass Thermometers, Still Well and Indicating," 1991 ISA Conference, Paper #91-0308.

15 Thermistors

T. J. CLAGGETT, R. W. WORRALL (1969, 1982) **B. G. LIPTÁK** (1993)

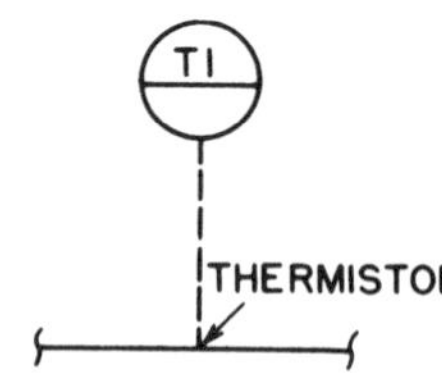

Flow Sheet Symbol

Range:	Metallic oxide thermistors are available from −150 to 600°F (−101 to 316°C); sintered metal and carbon thermistors are available down to −400°F (−240°C)
Span:	Can be as narrow as 2°F (1°C), which can result in extremely high sensitivity
Differential Temperature Measurement:	By connecting two thermistor sensors to different halves of the same bridge, thousandths of a degree differences can be measured
Sensor Resistances:	From 100 to 500,000 ohms. Standard units are about 5,000 ohms. If the resistance of a thermistor element is 5000 ohms at room temperature it will drop to about 20 ohms at 300°C (572°F) and will rise to 200,000 ohms if the temperature drops to −50°C (−58°F)
Linearity:	Nonlinear
Sensitivity:	Typically 50 to 500 ohms/degree, which translates into 1% to 10% of sensor resistance change per degree temperature. Most industrial units have a sensitivity of about 4%/°C. The sensitivity of Moxies is much greater
Inaccurary:	Generally between 0.1 and 1°F (0.06 and 0.6°C). Because spans can be extremely narrow, the error with this sensor (if calibrated) can be as low as 0.02°F (0.01°C), but because its long-term stability is low and because of self-heating effects, long-range accuracy is much lower
Cost:	The unpackaged elements can be obtained for a few dollars; when packaged for process industry services, their costs are similar to those of RTD-based sensors or transmitters
Partial List of Suppliers:	Analogic Corp.; Atkins Technical Inc.; Ball Corp.; Barber Colman Co.; Beckman Industrial Corp.; Conax Buffalo Corp.; Dale Electronics Inc.; Digitec; Dowty Custom Electronics Inc.; Fenwall Electronics Inc.; Gordon Co.; Hart Scientific Inc.; Honeywell Inc.; Horiba Instruments; Hy-Cal Engineering; Instrulab Inc.; JMS Southeast Inc.; Kane-May Measuring Instruments; Keystone Carbon Co.; Mitsubishi International Corp.; Murata Erie North America Inc.; Omega Engineering Inc.; Oven Industries Inc.; Pyromation Inc.; Pyrometer Instrument Co.; Qualimetrics Inc.; Reotemp Instrument Corp.; Tegam Inc.; Thermalogic Corp.; Thermetrics-Maveric Corp.; Thermometrics Inc.; Wahl Instruments Inc.; Wilkerson Instrument Co.; Yellow Springs Instrument Co.

Thermistors are semiconductors made from specific mixtures of pure oxides of nickel, manganese, copper, cobalt, tin, uranium, zinc, iron, magnesium, titanium, and other metals sintered at temperatures above 1800°F (982°C). Their distinguishing characteristics are a high temperature coefficient and the fact that their resistance is a function of absolute temperature. Bell Laboratories manufactured them some 60 years ago, naming them from the term "*therm*ally sensitive re*sistors*." While their use as temperature sensors was recognized over 100 years prior to the Bell investigation, the sensors were unstable and not reproducible. Therefore, they were seldom used outside the laboratory.

Their temperature coefficient is usually negative, although it can be positive as well. In some cases, with the negative coefficient type, thermistor resistance decreases at the rate of over 3% for each °F temperature rise. This makes them very good for narrow span measurement, but rather hard to handle for wide span applications.

Figure 15a illustrates the relative behavior of thermocouples, RTDs, and thermistors. On the right side of the figure the characteristics of a particular thermistor are shown, which illustrates both its sensitivity and its nonlinearity.[1]

SENSOR TYPES

The Bell project resulted in the development of thermistors stable and reproducible enough to make their large-scale use worthwhile in telephone work around 1940. However, industry in general did not accept these sensors until the 1950s. Bad experiences with commercially available thermistors hampered their acceptance. Variances in resistance at a given temperature and in rate of change of temperature

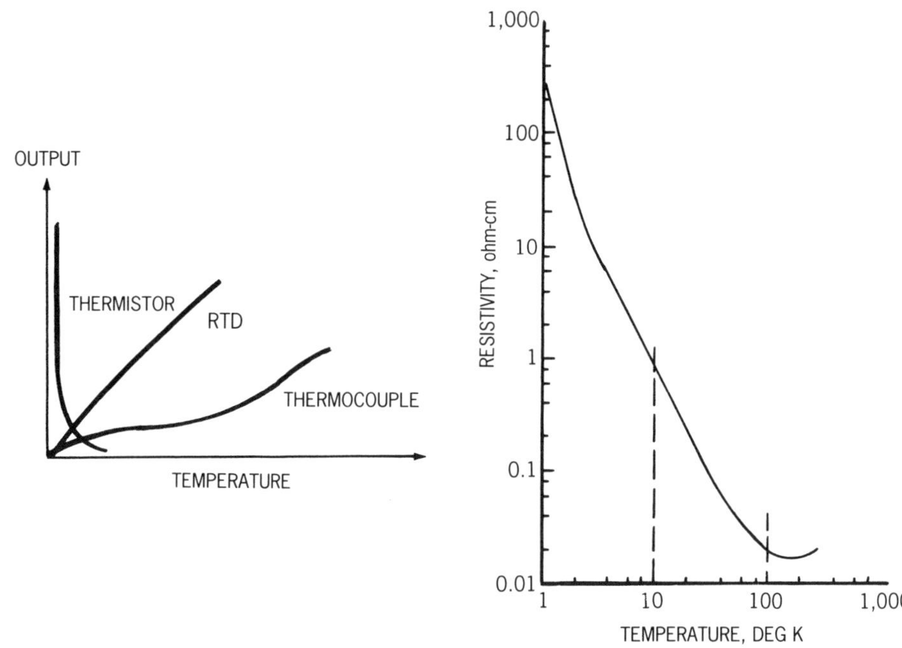

FIG. 15a
Thermistors are very sensitive and highly nonlinear.[1]

made individual calibration a requirement. Overcoming this difficulty, one manufacturer patented a process for interchangeable thermistors, resulting in production of probes interchangeable to 0.05°F (0.03°C).

A number of configurations are possible (Figure 15b). Most familiar is the bead type, which is usually glass-coated. However, thermistors can be made into washers, discs, or rods. They can also be encapsulated in plastic, cemented, soldered in bolts, encased in glass tubes or needles, or a variety of other forms.

These assemblies serve to support the sensor, protect against damage to the wires, direct flow across the unit uniformly, permit sealing of conduits or flowlines, and provide for easier handling.

Thermistors can be used in a number of applications other than temperature measurement. Some of these are voltage regulation, flowmeters, power level controls, gas analyzers, and vacuum detectors.

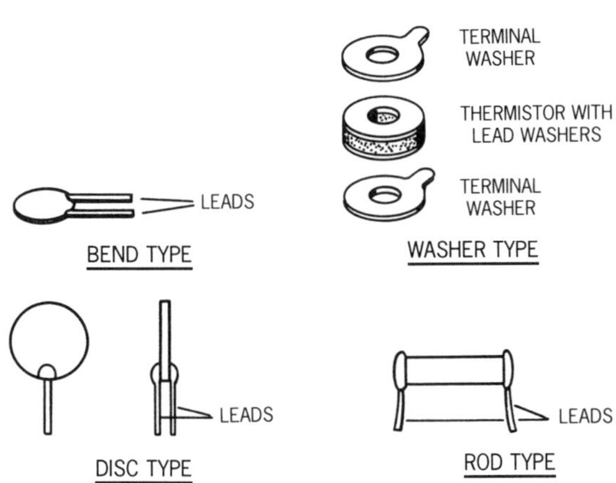

FIG. 15b
Variations in thermistor sensor packaging.

The I^2R temperature increase is a direct function of the dissipation constant in its mounting environment, and, unlike resistance thermometers or thermocouples, thermistor resistance values are varied by varying their composition to suit the temperature span, range, and sensitivity desired for a given application. If a small current flows through the thermistor, there is a negligible increase in its temperature due to this flow. The resistance can thus be measured and, since resistance is proportional to absolute temperature, the temperature can be inferred.

SELF-HEATING EFFECT

If the current flow is slowly increased, the heat generated within the sensor will gradually begin to raise its temperature above that of its environment. This in turn lowers its resistance and more current will flow. Eventually, the current input will reach a level where it is balanced by the heat output of the thermistor. At this point, sensor temperature would be 300 or 500°F (149 or 260°C) above ambient and sensor resistance would have dropped to a value approximately 10^{-3} times its value at the original low current.

For most temperature-measuring applications, self-heating is not a problem, since thermistor currents used are relatively low. If a current of higher magnitude is used, under fixed conditions, an offset allowance may be made for the self-heating effect. However, anything that affects the dissipation constant will change the offset. This could be the result of a number of things, including flow changes in the measured medium, changes in fluid composition, and the like.

The power dissipation factor of a thermistor element can

vary from a few microwatts to several watts per degree of resultant temperature rise. In industrial units a few degrees of self-heating can be expected for each milliwatt of resistive heating. As thermistors are made smaller and smaller in order to increase their response times and to lower their thermal shunting, the effect of self-heating also rises.

MEASURING BRIDGES

To measure temperature with a thermistor, a Wheatstone bridge circuit (Figure 15c) can be used as mentioned earlier. But, a simple circuit such as that shown in Figure 15d consisting of a battery, a sensor, and a microammeter can also be used. In such a circuit, the sensor will have a very high resistance. As long as the voltage is constant, the current flow will be determined only by changes in resistance of the thermistor. The sensor can be mounted fairly far from the meter and ordinary copper wire can be used for transmission. With a sensor resistance on the order of 10^5 ohms, meter temperature changes and leadwire resistances can be neglected.

Assuming that the thermistor changes its resistance by 4% (by 4000 ohms) for each degree of temperature change and assuming that the total lead wire used was 500 ft of 20 gauge copper wire having a resistance of 10 ohms, the leadwire error would amount to only 0.0025 degrees of temperature error. Even if the thermistor had a total resistance of only 5000 ohms, and therefore its 4% temperature coefficient were only 200 ohms/°C, the lead-wire error would still only be 10/200 = 0.05°C. For this reason lead-wire compensation is not a serious problem, and three- or four-wire bridges are not used with thermistors.

Locating the thermistor in one leg of a bridge circuit (Figure 15c) with a center zero galvanometer enables very narrow temperature spans to be displayed relatively inexpensively. Range depends upon galvanometer sensitivity and can be as low as 2°F (1°C).

Very accurate temperature measurements can be made with a differential circuit (Figure 13e). With two thermistors in different halves in a bridge, the unbalance will be determined by resistance difference caused by temperature difference of the sensors. With a high gain amplifier, differentials of 10^{-3} °F or °C can be measured.

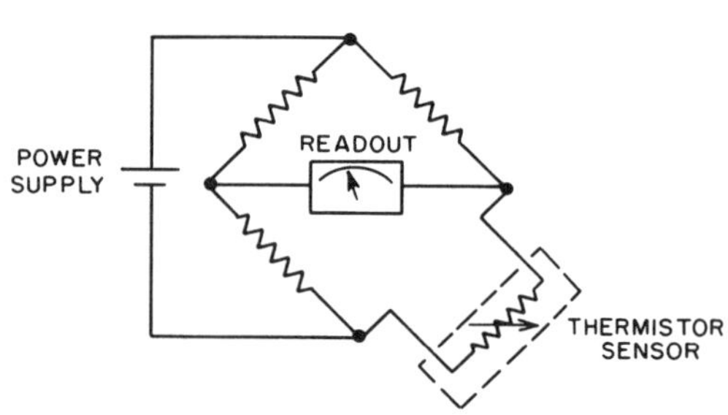

FIG. 15c
Wheatstone bridge.

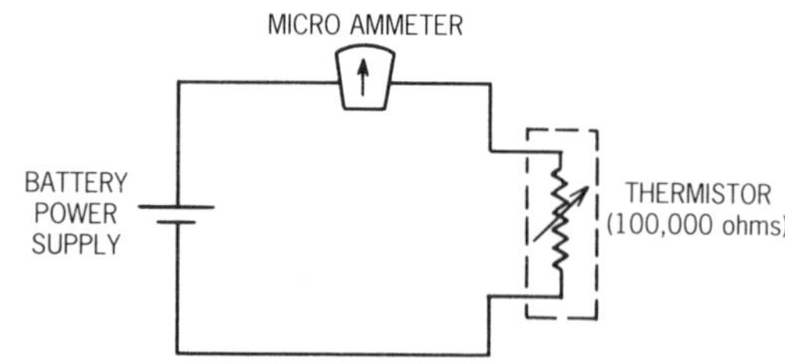

FIG. 15d
Microammeter readout.

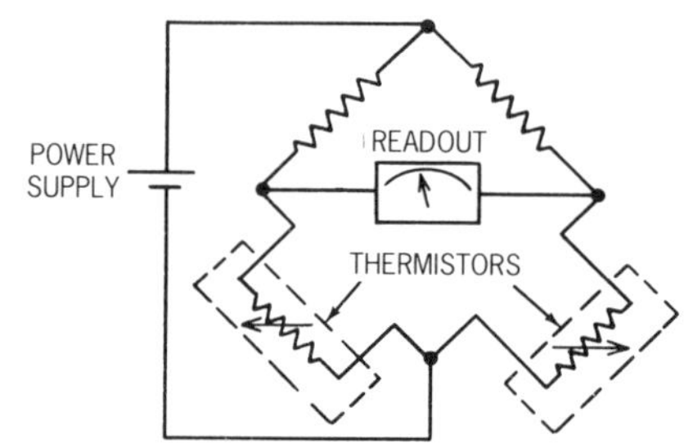

FIG. 15e
Differential measurement.

THERMISTORS COMBINED WITH RESISTORS

Combinations of thermistors and wire-wound resistors can be connected in networks (Figures 15f and 15g) to produce either a varying voltage or resistance which is linear with temperature. The basic equation for a divider network of R_1 and R_2 in series is $E_{out} = E_{in} R_1/(R_1 + R_2)$, where E_{out} is the voltage drop across R_1. When R_1 is a thermistor and E_{out} is plotted against temperature, the total curve is nonlinear and S-shaped. If R_1 is modified by the addition of another thermistor and resistor of proper values, linearity of the center section of the S curve can be extended to cover a relatively wide temperature range. This section then is considered to follow the general equation of a straight line. For the resistance mode, this would be $R_1 = MT + b$, where M is slope in ohms per degree, T is temperature in degrees F and b is value of R_1 when T = 0°F. The advantages of such a system are obvious.

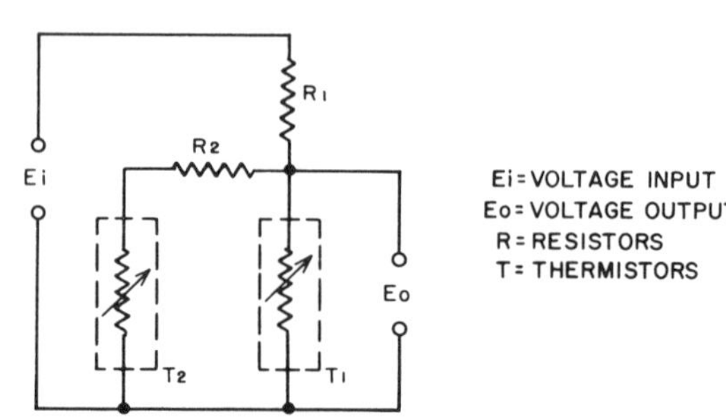

FIG. 15f
Two-thermistor network.

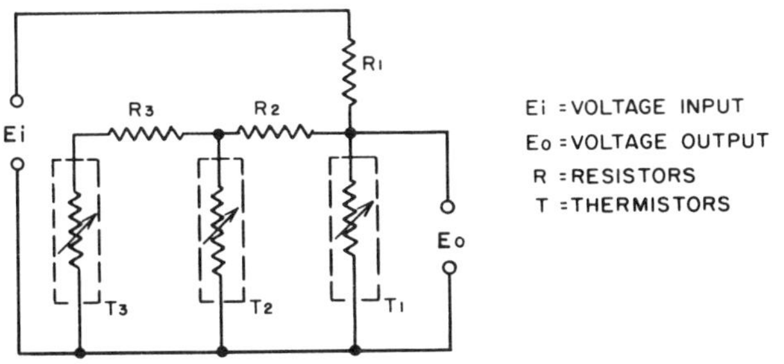

FIG. 15g
Three-thermistor network.

SPECIALIZED APPLICATIONS

Thermistors can be connected to a microprocessor-based memory element and packaged as a portable element (Figure 1e). These small micropacks can memorize the temperature history of batch reactor products or can travel with the process material through several steps in processing. When the temperature memory-pack unit is retracted from the process it can be plugged into a computer for interrogation. This allows the plant to store the temperature history of each batch. The micropack weighs less than 100 g, has a battery life of about 500 hours, and has a temperature range of −40 to 302°F (−40 to 150°C).

Another specialized application for thermistors is to use them as regenerative temperature switches. These units are produced through the use of thin film technology where a change in the the resistance of a metal oxide (Moxie) film is used as a temperature switch (Figure 15h). In the extended scale mode, Moxies behave like thermistors but exhibit a resistance change of up to 300%/°C. In the switching mode, they operate as thermally actuated switches. The switching temperature is solely determined by the programming voltage. On passing through their actuation temperature, Moxies abruptly switch to their low impedance state and latch.

CALIBRATION AND TESTING

The National Bureau of Standards has offered a limited calibration service for thermistors over a temperature range of approximately −150 to 200°F (−101 to 93°C). Calibrations on the order of ±.02°F (±.01°C) are the rule. Users have claimed better than ±.002°F (±.001°C) stability over a two-year period.

The following should be checked to test the operating parameters of thermistors:

1. Zero power resistance. This test is done under conditions that produce negligible heating of the sensor by test current. Most common sources of error are measurement of ambient temperature, self-heating error, thermocouple effects at junctions of dissimilar metals, and accuracy of test equipment. An ordinary ohmmeter is too crude. A five-dial Wheatstone bridge and a good mirror galvanometer will do the job.
2. Temperature coefficient of resistance. The rate of change of thermistor resistance versus temperature at the desired temperature.
3. Voltage developed across the thermistor under conditions of thermal equilibrium with a constant current. A current-limiting resistor should be used. Maximum current should not be exceeded, even for short periods. The sensor should not be moved to a medium of lower thermal conductivity during a test. The problem is similar to that of measuring voltage across ordinary resistors; the difference is that stabilization times are longer and small currents are used.
4. Time required for the thermistor to pass a certain current after the voltage is applied.
5. Dissipation constant measurements, which have been defined as the ratio at a certain temperature of a change in power dissipation in a thermistor to the resultant body temperature change.
6. Thermal time constant, which is the time for a 63.2% change from initial to final temperatures when subjected to a step change.

We have spoken mainly of negative coefficient thermistors, since they are most widely used. However, positive temperature coefficient thermistors are also used to protect motors from overheating, for very accurate measurements at high temperatures, for nonexpendable current limiters, and for flowmeters. In these applications, the positive increase in resistance with temperature and rapid change in sensitivity at switching temperatures are useful.

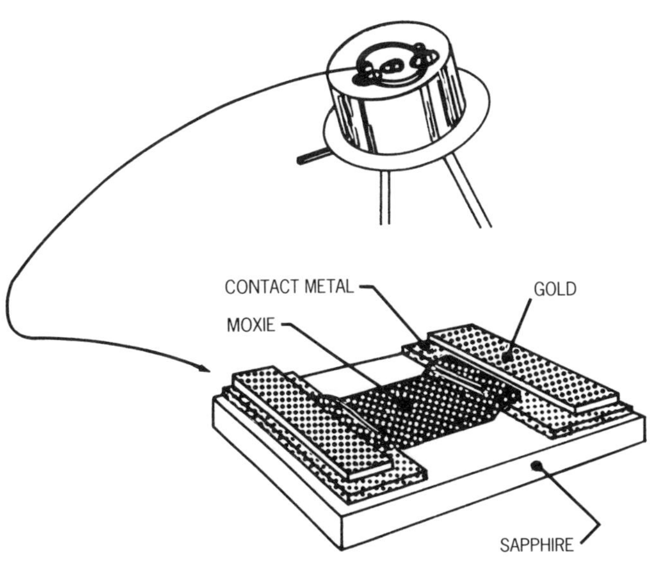

FIG. 15h
Thermally actuated switches. (Courtesy of Multi-State Devices Ltd.)

ADVANTAGES AND LIMITATIONS

Thermistors have the desirable characteristics of small size, narrow spans, fast response (their time constant can be under a second), and a very high sensitivity (about 4%/°C), which usually increases as the measured temperature drops. Thermistors do not need cold junction compensation because their resistance is a function of absolute temperature, and errors due to contact or lead-wire resistances are insignificant because of their relatively small values. Thermistors are available in a great variety of configurations, they are inexpensive, they are not affected by polarity, and their stability increases with age. They are the most sensitive differential temperature detectors available.

The disadvantages of thermistors are also many. They are neither interchangeable nor linear, although in modern data acquisition systems linearization is no problem. They are also fragile and are not suited for wide spans. Their high resistance necessitates the use of shielded power lines, filters, or DC voltage. One of the most serious limitations of thermistors is their lack of stability (drift and decalibration) at higher temperatures. This is because semiconductors in general change their composition and develop migration, diffusion, or other decomposition properties at high temperatures. Therefore, their use is limited to temperatures up to 600°F (316°C), and even at such temperatures extended exposure can cause drifts. (It should be noted that if a thermistor is operated at temperatures far below its maximum it can be a very stable device.) Thermistors also have a low temperature limit, because as the temperature drops their resistances rise to such levels that measuring it becomes difficult.

Reference

1. Nussbaum, A., ''Semiconductor Thermometers,'' *Measurement,* September 1961.

Bibliography

Droma, C.R., ''Thermistors for Temperature Measurements,'' in *Temperature, Its Measurement and Control in Science and Industry,* Vol. 3, New York: Van Nostrand Reinhold, 1962.

Green, C.B., ''Thermistors as Primary Temperature Elements,'' in *Process Instruments and Controls Handbook,* Consodine, D.M. (ed.), New York: McGraw-Hill, 1957.

Handbook of Thermistor Applications, Springfield, New Jersey: Victory Engineering Corp., 1968.

Hashemian, H.M., and Peterson, K.M., ''Assuring Accurate Temperature Measurement,'' *InTech,* October 1989.

Hormuth, G.A., ''Ways to Measure Temperature,'' *Control Engineering,* Reprint No. 948, 1971.

Instrumentation and Control Systems Engineering Handbook, TAB Books, 1978.

Kerin, T.W., and Katz, E.M., ''Temperature Measurement in the 1990s,'' *InTech,* August 1990.

Leewis, W., ''The International Temperature Scale of 1990,'' 1991 ISA Conference, Paper #91-0302.

Standard DIN 43760, ''Fundamental Values of Measuring Resistors and Resistance Thermometers,'' German Standards Committee (DNA).

16 Thermocouples

T. J. CLAGGETT, R. W. WORRALL (1969, 1982) **B. G. LIPTÁK** (1993)

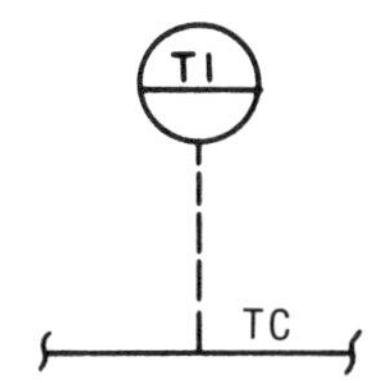

Flow Sheet Symbol

Linearity:	Thermocouples are nonlinear devices. They generate from 10 to 60 microvolts per °C temperature change
Temperature Ranges:	See Table 16l
Temperature Spans:	See Table 16m. Minimum recommended span is 25°C (45°F) and is available only in types E, J, K, and T. For types S and R the minimum span is 360°F (200°C)
Inaccuracy:	The total error is the sum of the thermocouple wire error (Table 16m) and the transmitter error. The transmitter error is usually 0.15% of span with standard and 0.05% of span with intelligent transmitters, or the sum of the cold junction (CJ) and the minimum absolute error (MAE), whichever is greater. With narrower spans the sum of the CJ + MAE error is likely to be the determining one and can be calculated as the sum of CJ = 0.25°C (0.45°F) and MAE = 0.2°C (0.36°F) for types T, J, E, and K; 0.6°C (1.08°F) for types R and S; and 0.8°C (1.44°F) for type B
Costs:	Laboratory probes which are detachable and are provided with handles can be obtained for $25 to $60. A thermocouple element in a stainless steel thermowell costs about $200, while a thermocouple element which is integral with a transmitter can range from $750 to $2000, depending on its features
Partial List of Suppliers:	ABB Kent-Taylor Inc.; Accutech; Action Instruments Co. Inc.; Alnor Instruments Co.; Altek Industries Corp.; Anritsu Meter Co. of America; ARI Industries Inc.; Athena Controls Inc.; Atkins Technical Inc.; Bailey Controls Co.; Barbeer Colman Co.; Barnant Co.; Beckman Industrial Corp.; Paul Beckman Co.; Burling Instruments Inc.; Burns Engineering Inc.; Capp Inc.; Celesco Transducer Products; Conax Buffalo Corp.; C-Temp Inc.; Delta M Corp.; Devar Inc.; Digitron Instrumentation Ltd.; Doric Scientific Corp.; Durex International Corp.; Electronic Development Laboratories Inc.; Enercorp Instruments Ltd.; The Eppley Laboratory Inc.; Fenwall Inc.; The Foxboro Co.; GIC Thermodynamics Inc.; Gordon Co.; Harrop Industries Inc.; Hart Scientific Inc.; Heimann Systems; Heraeus-Sensor; Honeywell Industrial Controls Div.; Hy-Cal Engineering; Industronics Service Co.; International Pressure Service Inc.; JMS Southeast Inc.; Jumo Process Control Inc.; Kaye Instruments Inc.; Labfacility Ltd.; Leeds & Northrup/General Signal; LFE Instruments; Love Controls Corp.; McFadden & Associates; Minco Products Inc.; Moore Industries International Inc.; Moore Products Co.; M-System Technology Inc.; Nanmac Corp.; National Basic Sensor; Newport Electronics Inc.; Norwich Aero Products Inc.; Omega Engineering; Omron Electronics Inc.; Pacer Industries; Projects Inc.; Promac Inc.; Pyro Inc.; Pyromation Inc.; Pyrometer Instrument Co.; RDF Corp.; Rochester Instrument Systems; Rosemount Inc.; S-Products Inc.; Sandelius Instruments Inc.; Service Associated Inc.; Simpson Electric Co.; Temco Products Co.; Temperature Measurement Systems Inc.; Temp-Pro Inc.; Thermetrics-Maveric Corp.; Thermo-Couple Products Co.; Thermo-Electric Co. Inc.; Thermometrics Corp.; Transmation Inc.; Tudor Technology Inc.; United Electric Controls Co.; Wahl Instruments Inc.; Weed Instrument Co.; Weksler Instrument Corp.; Westcon; West Instrument, Mark IV; YSI Inc. *Note*: The most popular thermocouple or RTD based transmitter suppliers are: Rosemount, Moore Industries, Accutech and Moore Products.

Imparting heat to the junction of two dissimilar metals causes a small continuous emf to be generated. One of the simplest of all temperature sensors, the thermocouple depends upon the principle known as the Seebeck Effect. T. J. Seebeck discovered this phenomenon in 1821, and in the ensuing years the thermocouple has become the most widely used electrical temperature sensor. The word is a combination of ''thermo'' for the heat requirement and ''couple'' denoting two junctions.

HEAT-OPERATED BATTERY

An ordinary thermocouple consists of two different kinds of wires, each of which must be made of a homogeneous metal or alloy. The wires are fastened together at one end to form a measuring junction, normally referred to as the hot junction, since a majority of the measurements are made above ambient temperatures. The free ends of the two wires are connected to the measuring instrument to form a closed path in

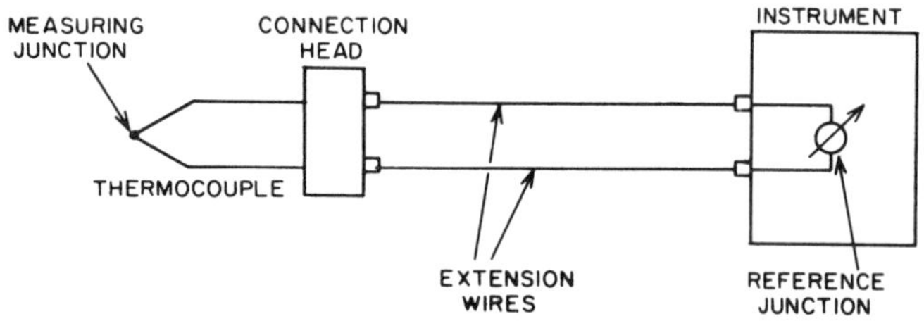

FIG. 16a
Thermocouple terminology.

which current can flow. After the thermocouple wires connect to the measuring instrument, the junction inside is designated as reference junction, or the cold junction (see Figure 16a).

The emf developed at wire junctions is a manifestation of the Peltier Effect and occurs at every junction of dissimilar metals within the measuring system. This effect involves the liberation or absorption of heat at the junction when a current flows across it. The resultant heating or cooling depends upon the direction of current flow. Applications of this principle are becoming increasingly useful in electric heating and refrigeration.

A second emf develops along the temperature gradient of a single homogeneous wire. This is the Thomson Effect. It is most important that each section of wire in a given circuit be homogeneous, since, with no change in the composition or physical properties along its length, the circuit emf depends only upon the metals employed and the temperature of their junction; circuit emfs are independent of both length and diameter of wires. Another reason for requiring homogeneous wire is that thermal emfs within a single strand passing from a warmer to a cooler area, or vice versa, will cancel one another.

Further, if both junctions of a homogeneous metal are held at the same temperature, the metal does not contribute to the net emf of a circuit. Since some thermocouples are made of extremely expensive metals, this fact can be used to cut costs by supplying copper extension wire for long runs.

It follows, then, that by holding temperatures constant at all junctions except one within a given circuit, we can measure temperature as a function of the hot junction temperature with respect to the cold junction temperature.

INTERPRETING THE GENERATED VOLTAGE

What the thermocouple reads is the difference between the temperatures of its measuring and reference junctions. (Actually, we cannot measure anything in the absolute; all we can do is to compare a known quantity against an unknown.) If we know what the reference temperature is, we can identify the unknown process temperature by measuring the voltage generated by the thermocouple:

$$\text{unknown temperature} = (\text{voltage}/(\text{Seebeck coefficient})) + \text{reference temperature} \quad 16(1)$$

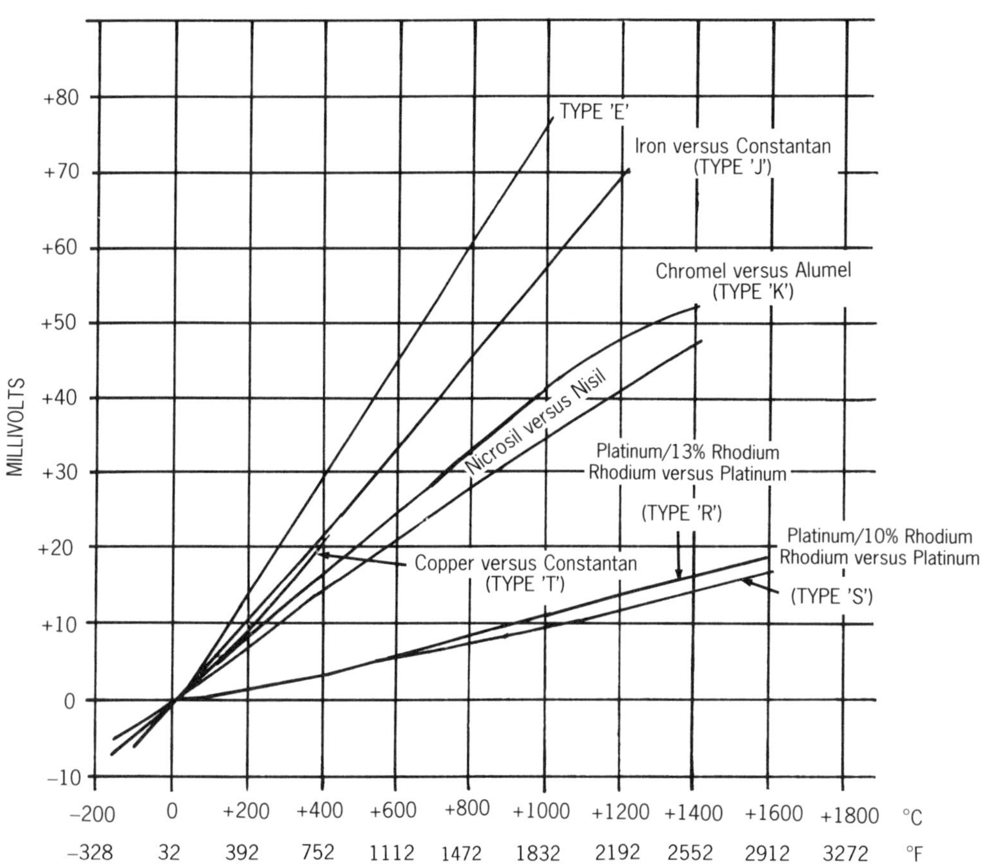

FIG. 16b
The millivoltage generated by thermocouples varies with wire materials and is a nonlinear function of temperature.

The process temperatures can be obtained from the voltage read by either going to a graph (Figure 16b) or, for more accuracy, by going to thermocouple tables that list the voltages corresponding to each temperature with each thermocouple type (such tables are provided at the end of this section). Unfortunately, the voltage-to-temperature relationship is not a straight-line function, and the Seebeck coefficient is not a constant (Figure 16c). For some thermocouples over certain temperature ranges, such as type K over the range 0 to 1000°C (32 to 1832°C), the Seebeck coefficient is relatively constant (about 40 microvolts/°C), but in general it changes with temperature. This in the past has resulted in unique scales for each type of thermocouple or in the need to use tables and curves to convert millivolts into temperature. These days the memory capability of the microprocessors has resolved all these problems, and what used to be tedious and time-consuming is now quick and easy. In short, the nonlinear nature of thermocouples is no longer a problem.

The same cannot be said about the weakness of the thermocouple signal. As shown in Figure 16c, a platinum thermocouple will generate only about 10 microvolts per degree Celsius. On the other hand, even the best industrial transmitters have a minimum span of 1 millivolt and a minimum absolute error of about 0.01 millivolt, which is 10 microvolts. Therefore it is difficult to obtain a measurement using industrial transmitter and platinum thermocouples, which would have less than a one degree Celsius error. This is usually acceptable for higher temperatures but is not acceptable when the temperature span is narrow. Therefore thermocouples are not recommended for narrow span or small temperature difference measurements.

THERMOCOUPLE TABLES

All thermocouple tables in this handbook (at the end of this section) are based upon a reference junction temperature of 32°F (0°C); therefore, direct conversion from the tables can be made only when an ice bath is used at the reference junction.

If it is not possible to maintain the reference junction temperature at 32°F (0°C), a correction factor must be applied to the millivolt values shown in the thermocouple tables. Note that the millivoltage produced by a given thermocouple is decreased when the temperature difference between the measuring junction and the reference junction is decreased. Correcting for reference junction temperatures other than 32°F (0°C) is described below.

Converting Millivoltage to Temperature

To apply the reference junction correction factor to a given potentiometer millivoltage reading, proceed as follows:

1. From the appropriate thermocouple table, obtain the millivoltage (based upon a 32°F R/J) corresponding to the actual temperature of the thermocouple reference junction.
2. ADD algebraically the value obtained in step 1, above, to the millivoltage read on the potentiometer.
3. The corrected millivoltage may then be converted into terms of temperature directly from the same table.

EXAMPLE 1

A potentiometer indicates a millivoltage of 13.019 mv when connected to a type T thermocouple, and it is desired to

TYPE	SEEBECK COEFFICIENT AT ROOM TEMPERATURE (μV/°C)
J	50
K	40
E	60
R	11
S	10
T	38

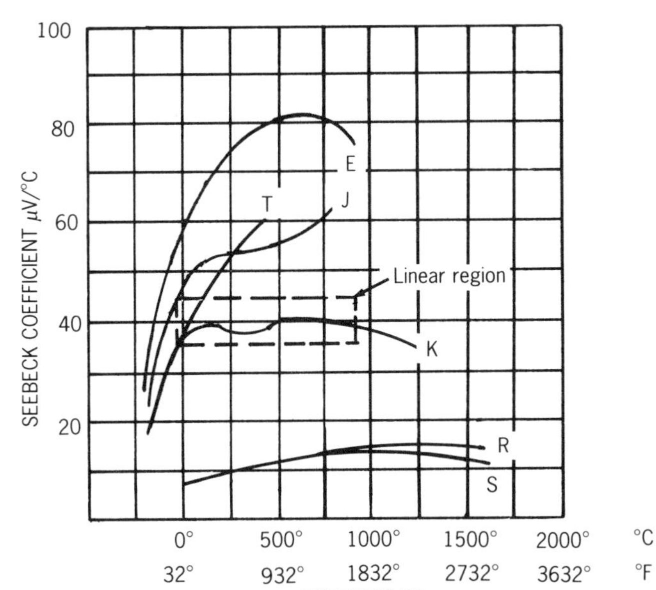

FIG. 16c
The Seebeck coefficient gives the amount of voltage generated (in microvolts) by a one degree change in temperature. The value of the Seebeck coefficient varies not only with thermocouple type but also with temperature.[1]

convert this value to its equivalent temperature. The actual thermocouple reference junction temperature, as determined by an accurate mercury-in-glass thermometer, is 68°F (20°C). Interpolating* from the type T table, 68°F = 0.787 mv, based upon a 32°F reference junction. Adding this value to the potentiometer reading, 13.019 + 0.787 = 13.806 mv, which is the corrected millivoltage based upon a 32°F reference junction. Interpolating from Table 16hh for type T table, 13.806 mv = 539°F (282°C).

EXAMPLE 2

A type T thermocouple under steady operating conditions causes a potentiometer reading of −3.357 mv. The actual thermocouple reference junction temperature is 70°F (21°C). From the type T table, 70°F = 0.832 mv based upon a reference junction of 32°F. Adding these two millivoltages algebraically, −3.357 + 0.832 = −2.525 mv. Interpolating,* −2.525 mv = −98°F (−72°C).

Converting Temperature to Millivoltage

To determine the proper millivolt input required to check the calibration of an instrument, proceed as follows:

1. From the appropriate table, obtain the millivoltage based upon a 32°F reference junction corresponding to the actual temperature at the input terminals of the instrument to be checked.
2. From the same table, obtain the millivoltage based upon a 32°F reference junction for the temperature to be checked.
3. Subtract algebraically the value obtained in step 1, above, from the value obtained in step 2.

EXAMPLE 1

It is desired to check the calibration of an instrument at 300°F (149°C). The instrument has a scale graduated in degrees Fahrenheit for a type T thermocouple. The actual temperature at the input terminals of the instrument to be checked, as determined by an accurate mercury-in-glass thermometer, is 70°F (21°C). From the type T table, 70°F = 0.832 mv and 300°F = 6.647 mv based upon a reference junction temperature of 32°F. Subtracting, the corrected millivoltage input required on the basis of a 70°F reference junction is 6.647 − 0.832 = 5.815 mv.

*To interpolate between two printed values, add algebraically to the smaller value a proportionate part of the difference between the two printed values, thus:

EXAMPLE 1 (Positive Temperature)

248°F = 245° + ³⁄₅(250° − 245°).

In terms of millivoltage from the type T table,

248°F = 5.147 + ³⁄₅(5.280 − 5.147)
= 5.147 + 0.0798 = 5.227 mv

EXAMPLE 2 (Negative Temperature)

−248°F = −245° + ³⁄₅(−250° − (−245°))

In terms of millivoltage from the type T table,

−248°F = −4.688 + ³⁄₅(−4.747 − (−4.668))
= −4.688 + ³⁄₅(−4.747 + 4.688)
= −4.688 − 0.0354 = −4.723 mv

EXAMPLE 2

It is desired to determine the correct millivoltage input required to check the calibration of an instrument at −200°F (−129°C). The instrument scale is graduated in degrees Fahrenheit for a type T thermocouple. The actual temperature at the input terminals of the instrument is 68°F (20°C). From the type T table, 68°F = 0.787 and −200°F = −4.111 mv based upon a 32°F reference junction. Subtracting algebraically, the corrected millivolt input on the basis of a 68°F reference junction is −4.111 − 0.787 = −4.898 mv.

THE LAWS OF INTERMEDIATE TEMPERATURES AND METALS

The law of intermediate temperatures states that the sum of the emf's generated by two thermocouples, one with its junctions at 32°F (0°C) and some reference temperature, the other with its junctions at the same reference temperature and the measured temperature, is equivalent to that produced by a single thermocouple with its junctions at 32°F (0°C) and the measured temperature.

This is represented in Figure 16d where the measured temperature is 700°F (371°C). By adding an emf equal to that produced by thermocouple A in Figure 16d (with its junctions at 32°F (0°C) and the reference temperature) to that of thermocouple B, a total emf equivalent to that generated by the hypothetical thermocouple C results. In most pyrometers, this is done by a temperature-sensitive resistor, which measures the variations in reference junction temperature caused by ambient conditions, and automatically provides the necessary emf by means of a voltage drop

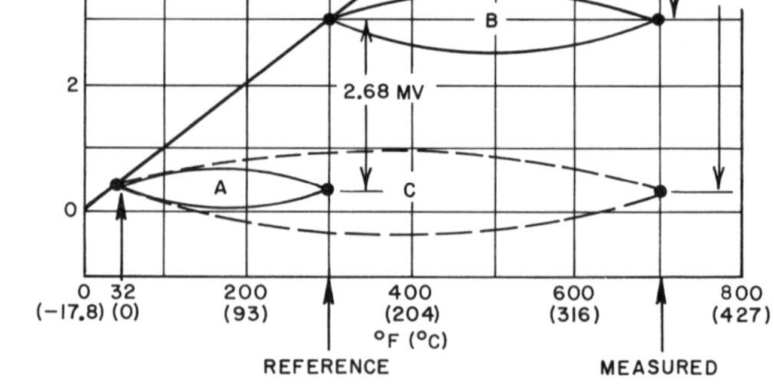

FIG. 16d

According to the law of intermediate temperatures, the emf of thermocouple A plus the emf of thermocouple B is equal to the emf of thermocouple C.

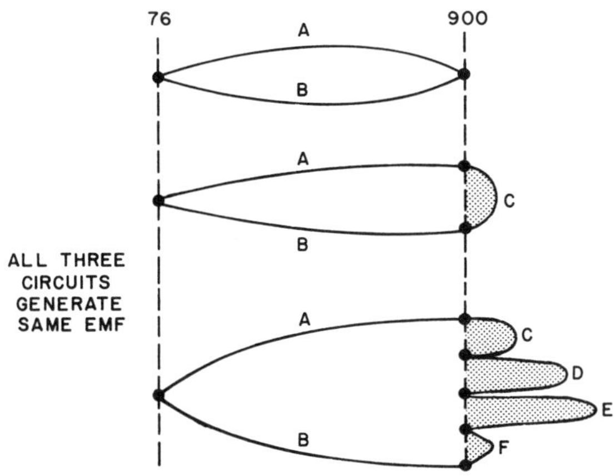

FIG. 16e
No harmful effect is caused by introducing any number of metals at a thermocouple junction if all connections are at the same temperature.

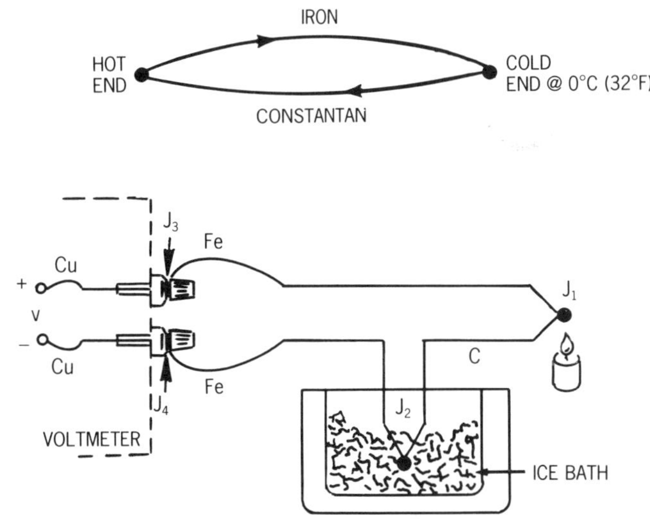

FIG. 16f
When an iron-constantan thermocouple measures the process (J_1) and an identical iron-constantan thermocouple reference junction (J_2) is placed in an ice bath, the connection to the readout voltmeter results in two added junctions, J_3 and J_4.[1]

produced across it. Thus, the instrument calibration becomes independent of reference temperature variations.

The law of intermediate metals states that the introduction of a third metal into the circuit will have no effect upon the emf generated so long as the junctions of the third metal with the other two are at the same temperature. Any number of different metals can be introduced, providing all the junctions are at the same temperature. Thus, in Figure 16e the circuits shown all generate the same electromotive force, even though the second and third circuit diagrams show materials C, D, E, and F inserted between A and B.

COLD JUNCTION COMPENSATION

When a readout device is employed, it converts the emf produced by the temperature difference between the hot and cold junctions to record or otherwise display the temperature of the hot junction. To prevent errors due to secondary emf's produced by variations of temperature at the cold junction and within the readout device, these emfs must be compensated for. One method is to hold the cold junction at a constant temperature, as can be done in laboratories with an ice bath (Figure 16f). An oven can also be used, although keeping an oven temperature constant presents another set of problems.

Neither an ice bath nor an oven reference is practical in an industrial environment. In the temperature transmitters used in the process industry, the ice bath reference must be replaced by a variable ambient reference junction. This is achieved by making two changes to Figure 16f. The first change is to insert a short copper wire between both voltmeter terminals and the thermocouple leads and to place these new junctions on an isothermal block (Figure 16g). This change eliminates the junctions J_3 and J_4 shown in Figure 16f because in Figure 16g copper is joined to copper at these points. By placing the new J_3 and J_4 junctions on an isothermal block, as shown in Figure 16g, their effects cancel out as they are in opposition to each other and are at the same temperature.

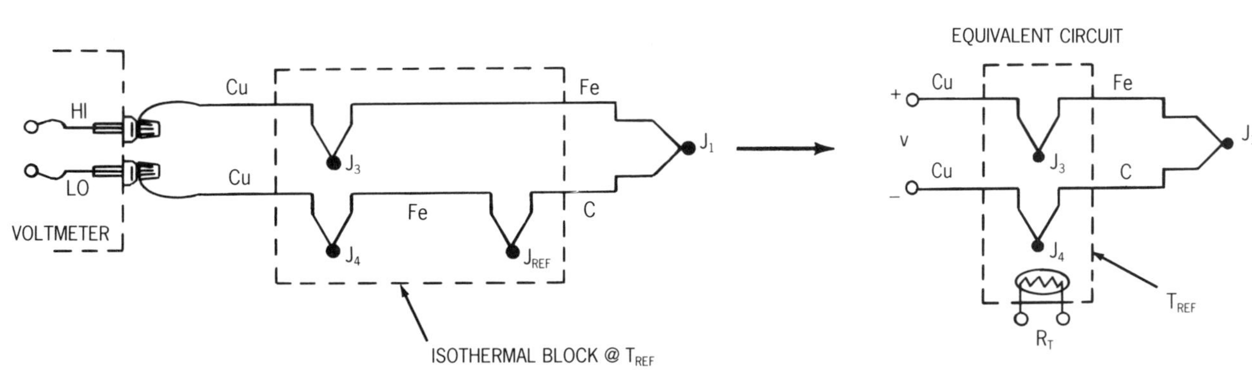

FIG. 16g
By the using an isothermal block and by inserting copper wires at the voltmeter terminal, an equivalent circuit is arrived at which does not have an ice bath reference.[1]

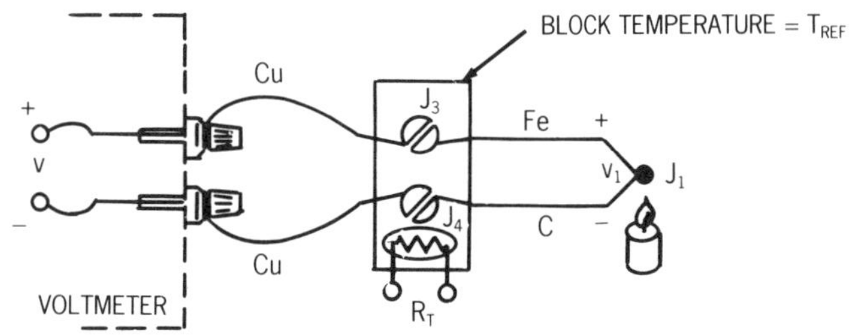

FIG. 16h
The industrial equivalent of the ice bath reference shown in Figure 16f.[1]

The second change was to place the reference junction not in an ice bath (Figure 16f) but on the isothermal block. From the law of intermediate metals (Figure 16e) we know that when junctions in series are at the same temperature, their number makes no difference. Therefore J_4 and J_{REF} in Figure 16g can be replaced by J_{REF} only.

Figure 16h shows the "software compensation" of the reference junction. Here the voltmeter reads the equivalent of thermocouple B in Figure 16d, while the thermometer R_T reads the actual reference temperature of the isothermal block. The thermometer used to measure T_{REF} can be a thermistor, an RTD, or an integrated circuit transistor. Once T_{REF} is accurately measured (usually within 0.25°C or 0.45°F), the associated software determines the corresponding millivoltage which a thermocouple *would have generated* if its hot junction were at T_{REF} and its cold junction were in an ice bath (thermocouple A in Figure 16d). The sum of A and B then represents the measured process temperature (referenced to ice) and can be looked up in the type of tables that are provided at the end of this section.

One might ask, Why use a thermocouple at all, if in order to measure the reference temperature another thermometer is needed? The answer to that question is simple: Do not use a thermocouple if another thermometer can measure the temperature. Unfortunately, the devices listed as good candidates to measure the ambient T_{REF} are not suited for the measurement of high temperatures or wide spans. For these applications the natural candidate is the thermocouple, and when it is used, it must be compensated.

The cost and complexity of software compensation is reduced when many thermocouples are multiplexed into the same readout device (Figure 16i). In that case, a large number of thermocouples can be terminated on the same isothermal block and a single thermometer can serve to provide software compensation for all. The disadvantages of thermocouple multiplexing include the necessity of transporting the low-level signals over some distance to the multiplexer (which can introduce common and normal mode noise, discussed later), the relatively high costs of thermocouple lead wires, and the added error due to the variable contact resistances of the multiplexer. Even with gold-plated contacts there will be at least one microvolt drop through the contacts, which in case of platinum thermocouples corresponds to an error of about 0.2°F (0.1°C).

Hardware Compensation

Prior to the advent of microprocessors and the associated software compensation of thermocouples, hardware compensation was used. Hardware compensation can be viewed as inserting a battery that cancels the offset voltage produced by the reference junction. These commercially available circuits provided an electronic ice point reference for one or many thermocouples. Their main advantage relative to software compensation is speed, because the computation time is eliminated. The main disadvantage of hardware compensation is that each gain resistor is suited to compensate only a particular type of thermocouple, while software compensation accepts any thermocouple. In practice, hardware compensation is usually accomplished by using resistors whose combined temperature resistance coefficient curves match those of the voltage-temperature curves produced by the reference junctions, canceling any variations in the cold junction temperature.

The Seebeck emf can be measured with either a millivoltmeter (Figure 16j) or a potentiometer (Figure 16k) circuit. We should remember that the thermocouple measures only the difference between its reference and hot junc-

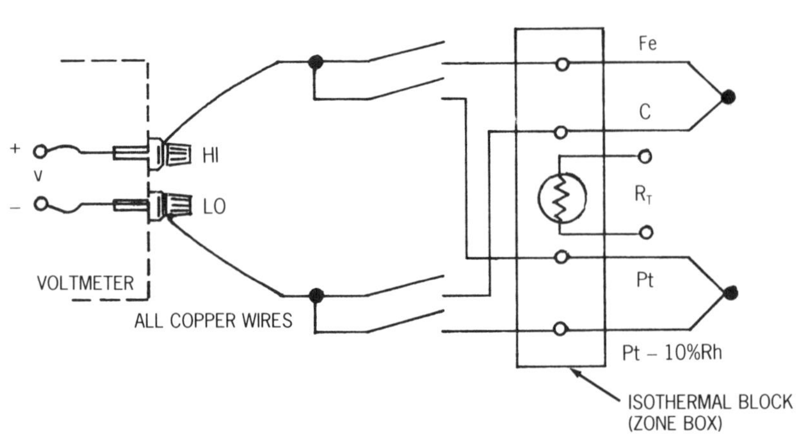

FIG. 16i
When many thermocouples are multiplexed a single reference thermometer is sufficient to provide software compensation.[1]

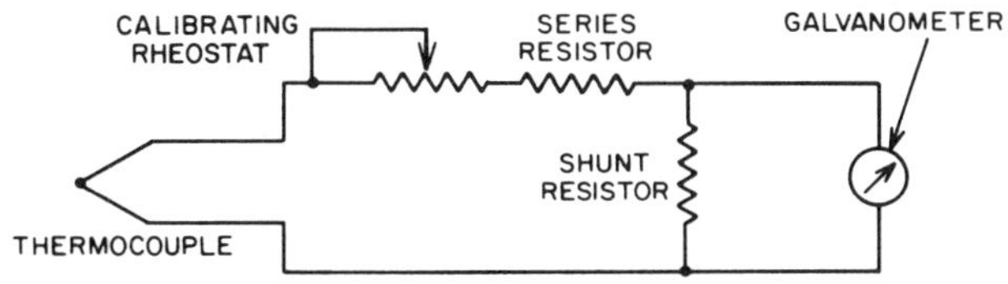

FIG. 16j
Millivoltmeter circuit.

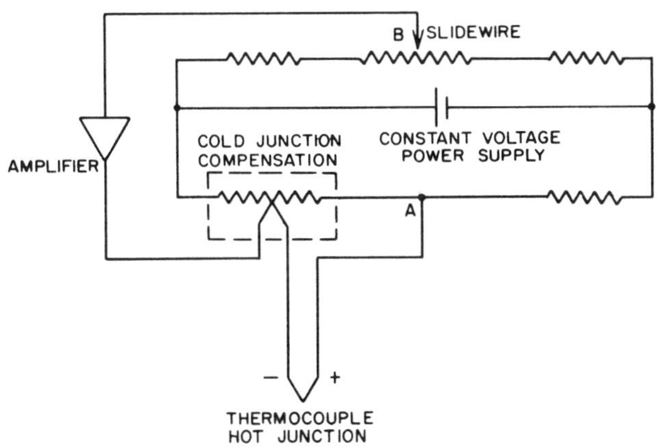

FIG. 16k
Potentiometer circuit.

tions. How closely it matches the accepted emf curve has a bearing on accuracy. Emf tables are usually based upon 32°F (0°C) reference temperatures for convenience.

To relieve the control engineer of the problem of compensating for temperature instability at the reference junction, a copper or nickel resistor can be placed in a bridge so that the thermocouple emf is opposed by an emf corresponding to the required ambient temperature correction. Operating on the null-balance principle, the resulting potentiometer (Figure 16k) tends to reduce any voltage difference between points A and B to zero.

THERMOCOUPLE TYPES

Based on possible combinations of metals there could be countless numbers of thermocouples; but, as a matter of fact there are relatively few (see Tables 16l and 16m). Things that determine a metal's usefulness in thermocouple wire include:

1. Melting points
2. Reaction to various atmospheres
3. Thermoelectric output in combination
4. Electrical conductance
5. Stability
6. Repeatability
7. Cost
8. Ease of handling and fabrication

The platinum-platinum 90%/rhodium 10% (ISA type S) thermocouple is most important, for it is used to define the International Temperature Scale between 1166.9°F (630.5°C), the point at which antimony freezes, and 1945.4°F (1063°C), the gold point. This thermocouple is not limited to the above range. It can be used from about 300 to 3215°F (150 to 1768°C) with excellent results. Industrial thermocouples ("Special" in Table 16m) of this material will match the standard calibration curve to better than ±.25%.

Several other thermocouples are commonly used (see Tables 16l and 16m for a comparison). Platinum-platinum 13% rhodium (type R) and platinum 30% rhodium-platinum 6% rhodium (type B) are recommended for use in oxidizing atmospheres. They are relatively easily contaminated in other atmospheres.

Copper-constantan (type T) can be used in either oxidizing or reducing atmospheres. Thermocouples of this type exhibit a high resistance to corrosion from moisture, provide a relatively linear emf output, and are good from the medium to the very low temperature range.

Iron-constantan (type J) can be used in reducing atmospheres. These thermocouples provide a very nearly linear emf output. They are the least expensive commercially available type.

Chromel-alumel (type K) can be used in oxidizing atmospheres. They are the most linear thermocouple in general use.

Chromel-constantan (type E) thermocouples provide the highest emf per degree of temperature change. However, they also tend to drift more than the others. They can be used in oxidizing atmospheres.

Tungsten-tungsten 26% rhenium thermocouples can be used to measure the highest temperatures. They cannot be used in oxidizing atmospheres and are brittle and hard to handle. They are used in a vacuum or in a clean inert gas.

A new base-metal thermocouple is designated type N (Nicrosil versus Nisil). It provides stability as good as the more expensive noble metal thermocouples up to about 2200°F (1204°C), where type K starts to become unstable. The stability of type N thermocouples is due to increased percentages of chromium, silicon, and magnesium.

THERMOCOUPLE CONSTRUCTION AND PROTECTION

A thermocouple is only as accurate as the wire from which it is made. Therefore, it is common practice for best accuracy to make all thermocouples from the same coil of wire. This assumes uniformity of the wire. Most manufacturers offer either standard or special calibrations which imply more care in selection of wire, handling, and manufacturing. But the careful selection of materials, proper construction, installation, and handling alone will not maintain highest accuracy; an adequate checking program is a must.

In order to protect the thermocouple wire, it is usually covered by thermal insulation and a sheath for mechanical protection. The purpose of this design is to expose only the measuring junction of the thermocouple to the temperature

TABLE 16I

Thermocouple Comparison Table

ISA Type Designation	Positive Wire (Numbers = Percentages)	Negative Wire (Numbers = Percentages)	Millivolts per °F	Recommended Range Limits Temp °F* Min.	Recommended Range Limits Temp °F* Max.	Scale Linearity	Atmosphere Environment Recommended	Favorable Points	Less Favorable Points
B	Pt70-RH30	Pt94-Rh6	.0003–.006	32	3380	Same as for type R couple	Inert or slow oxidizing	—	—
E	Chromel	Constantan	.015–.042	−300	1800	Good	Oxidizing	Highest emf/°F	Larger drift than other base metal couples
J	Iron	Constantan	.014–.035	32	1500	Good; nearly linear from 300–800	Reducing	Most economical	Becomes brittle below 32°F
K	Chromel	Alumel	.009–.024	−300	2300	Good; most linear of all TCs	Oxidizing	Most linear	More expensive than T or J
R	Pt87-Rh13	Platinum	.003–.008	32	3000	Good at high temps. Poor below 1000°F	Oxidizing	Small size, fast response	More expensive than type K
S	Pt90-Rh10	Platinum	.003–.007	32	3200	Same as R	Oxidizing	Same as R	More expensive than type K
T	Copper	Constantan	.008–.035	−300	750	Good but crowded at low end	Oxidizing or reducing	Good resis. to corrosion from moisture	Limited temp.
Y	Iron	Constantan	.022–.033	−200	1800	About same as type J	Reducing	—	Not industrial standard
—	Tungsten	W74-Re26	.001–.012	0	4200	Same as R	Inert or vacuum	High temp.	Brittle, hard to handle, expensive
—	W94-Re6	W74-Re26	.001–.010	0	4200	Same as R	Inert or vacuum	Same as above	Slightly less brittle than above
—	Copper	Gold-Cobalt	.0005–.025	−450	0	Reasonable above 60°K	—	Good output at very low temp.	Expensive lab.-type TC
—	Ir40-Rh60	Iridium	.001–.004	0	3800	Same as R	Inert	—	Brittle, expensive

$$*°C = \frac{°F - 32}{1.8}$$

TABLE 16m
Thermocouple Errors and Spans

TC Type	Measured Temperature Range in °F*	TC Wire Errors for Wires of Different Qualities*		Transmitter Error is Additional and in the Case of "Smart" Units Is ±0.05% of Span or Value Given Below, Whichever Is Larger*	Recommended Span Limits*	
		Standard	Special		Min.	Max.
B	32–3380	NA	NA	±1.89°F	63°F	2020°F
E	32–600 600–1600	±3°F ±0.5%	— —	±0.81°F	45°F	2100°F
J	32–530 530–1400	±4°F ±0.75%	±2°F ±0.375%	±0.81°F	45°F	2500°F
K	32–530 530–2300	±4°F ±0.75%	±2°F ±0.375%	±0.81°F	45°F	2750°F
R	32–1000 1000–2700	±5°F ±0.5%	±2.5°F ±0.25%	±1.53°F	360°F	2950°F
S	32–1000 1000–2700	±5°F ±0.5%	±2.5°F ±0.25%	±1.53°F	360°F	2900°F
T	−300 to −75 −150 to −75 −75–200 200–700	— ±2% ±1.5°F ±0.75%	±1% ±1% ±0.75°F ±0.375%	±0.81°F	45°F	1025°F
N	32–530 530–2300	±4°F ±0.75%	±2°F ±0.4%	NA	NA	NA

$$*°C = \frac{°F - 32}{1.8}$$

of the process. This can be achieved in three different ways (see Figure 16n). The exposed thermocouple junction gives the best speed of response; the time constant can be less than a second with small (down to 0.01 mm diameter) thermocouples. Their main limitation is that the process materials must not be corrosive to the thermocouple wires.

In the ungrounded junction design the thermocouple wire is physically insulated from the sheath by insulation material (usually magnesium oxide powder). These designs can be used in corrosive processes, but their speed of response is slow. The grounded junction design is also protected from the corrosive process, but its thermal time constant is shorter (by a few seconds, depending on mass).

The thermocouple extension wire is usually insulated with Teflon, PVC, nylon, rubber, asbestos, or fiberglass. For higher temperatures refrasil or nextel are recommended. Teflon is used when the thermocouple extension wire must be submerged under water or if resistance to solvent, corrosion, flame, or humidity is critical. Individually insulated duplex wires are usually provided with a protective outer jacket which can be wrapped, extruded, or stranded. The extension wire to be used for types E, J, K, and T thermocouples are designated as EX, JX, KX, and TX extension wires and should extend all the way to the cold junction of the loop.

With connections correctly made, copper extension wire can be used over long distances. However, it is recommended that iron-constantan and copper-constantan always be used with lead wire of the same material. To guard against mistakes in connection, industry practice is to color-code the wires, with the *negative* lead always *red.* Smaller gauge wire provides faster response, but heavier gauge wires last longer and resist contamination or deterioration at high temperatures.

The sheath material can be 304 stainless steel if the process temperature is under 1650°F (900°C) and the process is not highly corrosive. In furnaces that operate at up to 2100°F (1150°C), Inconel 600 sheathing is recommended if the atmosphere is oxidizing and there is no sulfur in the atmosphere. Platinum-rhodium alloy sheaths are used up to 3000°F (1650°C) in oxidizing furnaces if no silica or halogens are present. Molybdenum sheaths can be used up to 4000°F (2205°C) to detect molten metal or glass temperatures, but only in oxygen-free vacuum or inert-gas-filled processes. Tantalum sheaths can be used up to 4500°F (2482°C), but only in reducing or noble gas atmospheres

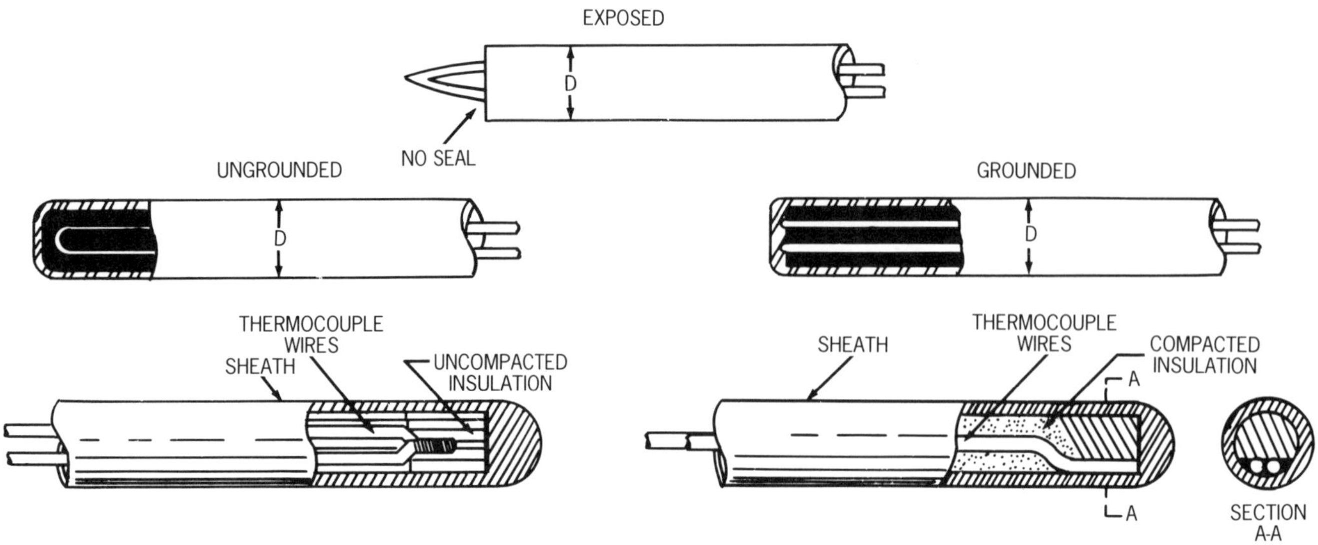

FIG. 16n
Thermocouple measuring junction designs. (Courtesy of ARI Industries Inc.)

where no oxygen is present. The sheath is usually strong enough to stand up to high pressure (up to 50,000 PSIG, or 3450 bars), but it is usually not used without a thermowell, because the user wants to be able to take out the thermocouple without opening up the process. Figure 16o illustrates some high-speed thermocouple assemblies without thermowells.

For calibrating thermocouples, depending upon the application, various procedures are used. Primary standard thermocouples of Pt-Pt, 10% Rh can be calibrated by the National Bureau of Standards to fixed points on the International Practical Temperature Scale. However, these thermocouples must be handled carefully to retain their accuracy. Thermocouples can be supplied by most major manufacturers calibrated against Primary Standard thermocouples kept in their own metrology laboratories. Secondary reference thermocouples for in-plant use are usually made of base metal. Comparison of these against the Primary Standard thermocouple is accomplished by placing them in close contact in a checking furnace. Users normally check their ordinary thermocouples against these secondary standards.

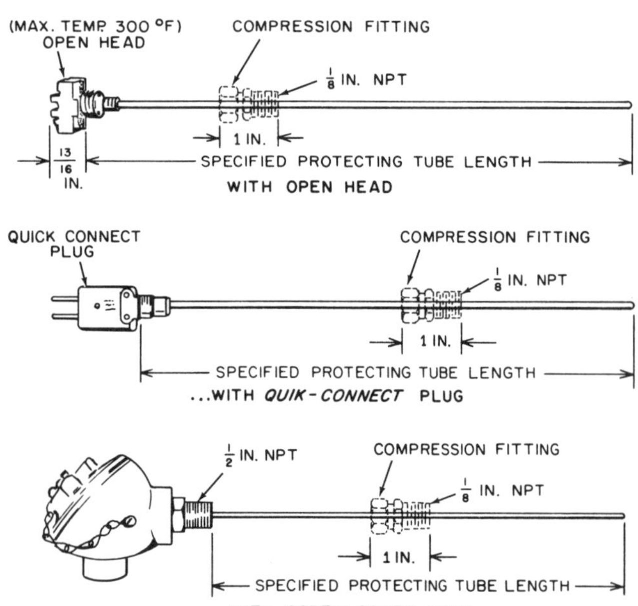

FIG. 16o
High-speed small O.D. thermocouple assemblies with stainless steel protecting sheath.

Thermowells

Protecting tubes or wells are supplied (Figure 16p) to protect thermocouples from harmful atmospheres, corrosive fluids, or mechanical damage; to support the thermocouple; or to permit thermocouple entry into a pressurized system. These tend to reduce the speed of response of the thermocouple, so small-mass, thin-wall, or needle-type installations are supplied where feasible (Figure 16o). Disposable-tip thermocouples are supplied in furnace applications (Figure 16q). They can also be peened or welded into a tube or tank well. Their low cost makes it feasible to place them in concrete beams while curing or to use them in other single-time operations.

When it is desirable to maximize the speed of response of the measurement, but it is also necessary to periodically remove the sensor, the bare (sheathed) thermocouple can be removed through a stuffing box and gate valve combination (Figure 16r). Most thermocouples are installed in a protecting well. In Figure 16s the R dimension is the "immersion length" while the U dimension is the "insertion length" of the well. R should be at least 10 times the diameter of the protective tube (sheath) diameter of the thermocouple. The sheath diameter of different thermocouples can range from 0.04 to 0.84 in. (1 to 21 mm), while the thermocouple wires

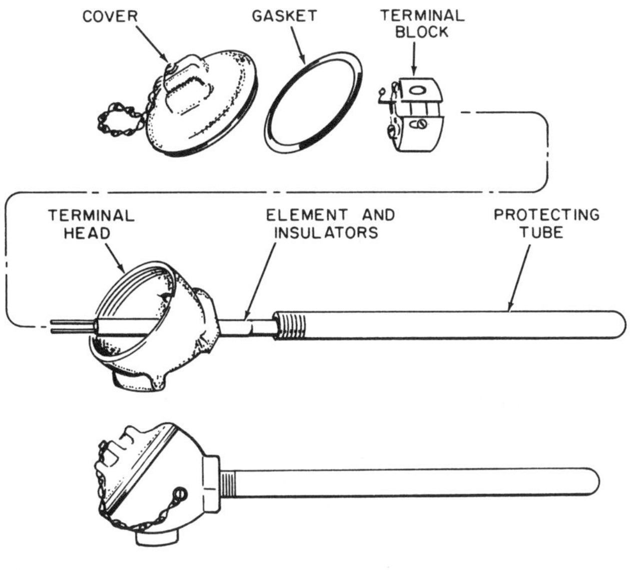

FIG. 16p
Exploded view of thermocouple assembly and protecting tube (top)*; complete assembly with protecting tube* (bottom).

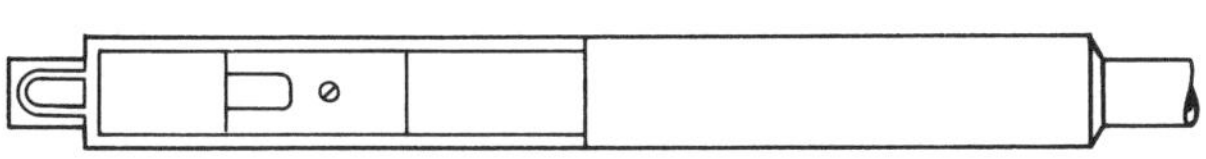

FIG. 16q
Molten steel expendable thermocouple.

themselves can range from gauge #36 to #8. The well can be inserted perpendicularly into the pipeline if R is not much more than the inside radius of the pipe. Otherwise, it should be inserted at a 90-degree bend in the line. The most often used well materials are 304 and 316 stainless steel, which are usable up to 1200°F (649°C). At higher temperatures ceramic thermowells are used because metallic ones start to ''droop'' (bend by gravity). High-purity alumina can be used up to 2200°F (1200°C); the same limit holds for mullite, but this material is not recommended for use with platinum thermocouples as it contains impurities which can

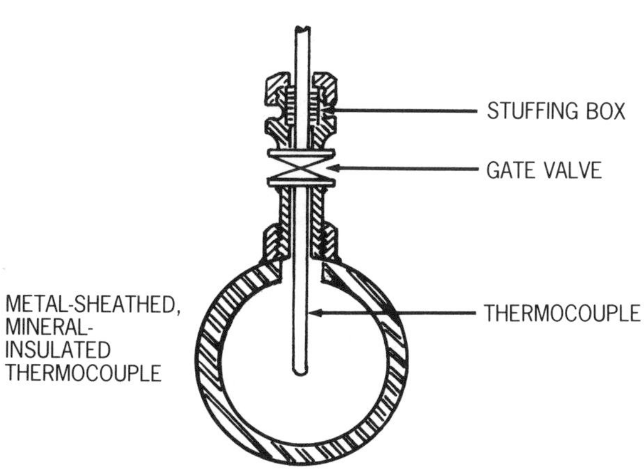

FIG. 16r
Installation of thermocouples without thermowells.[2]

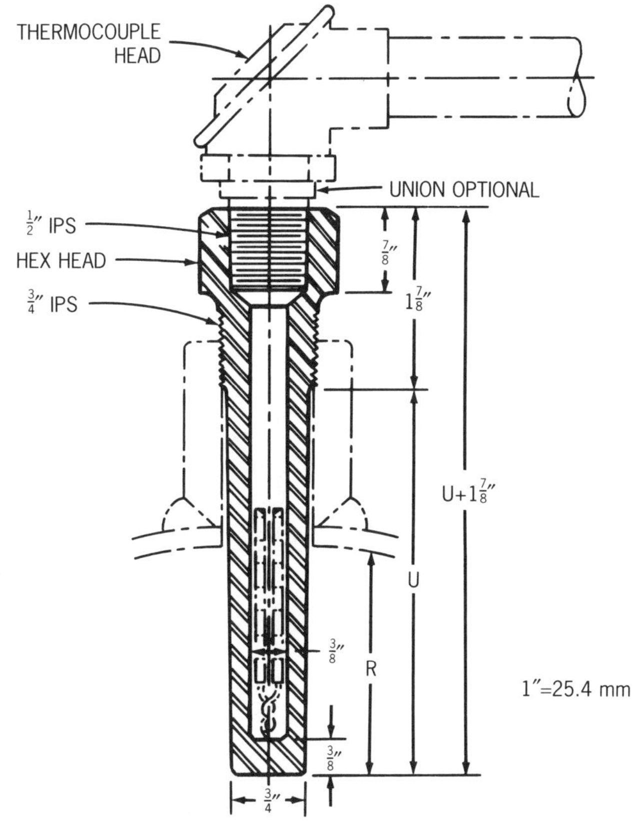

FIG. 16s
Screwed thermowell installation.[2]

contaminate platinum. The thermowell can have screwed connections (Figure 16s) or, if frequent inspection is required or if the well is glass-coated, it can be flanged.

When the surface temperature of tubes is to be measured, the thermocouple must be shielded from furnace radiation. The thermocouple can be attached to the heater tube surface by being furnished with stainless steel welding pads (Figure 16t, lower part) or by the use of thermocouple attachment blocks (Figure 16t, upper part). The multiple holes in these blocks allow for spare thermocouple elements for quick replacement.

MULTIPLE THERMOCOUPLE CONFIGURATIONS

The reason for inserting several thermocouples within the same thermowell can be to obtain a temperature profile over some distance. In this case each thermocouple junction is located at a different distance from the tip. In order for such sensors to detect the temperature outside the well (and not the air temperature inside it), it is essential that good physical contact be made between the thermocouple junction and the metallic surface of the inside of the well.

Thermocouples can be connected in parallel to provide the average temperature in a system (Figure 16u). They can also be used to measure the difference between two temperatures (Figure 16v). A single thermocouple can be utilized,

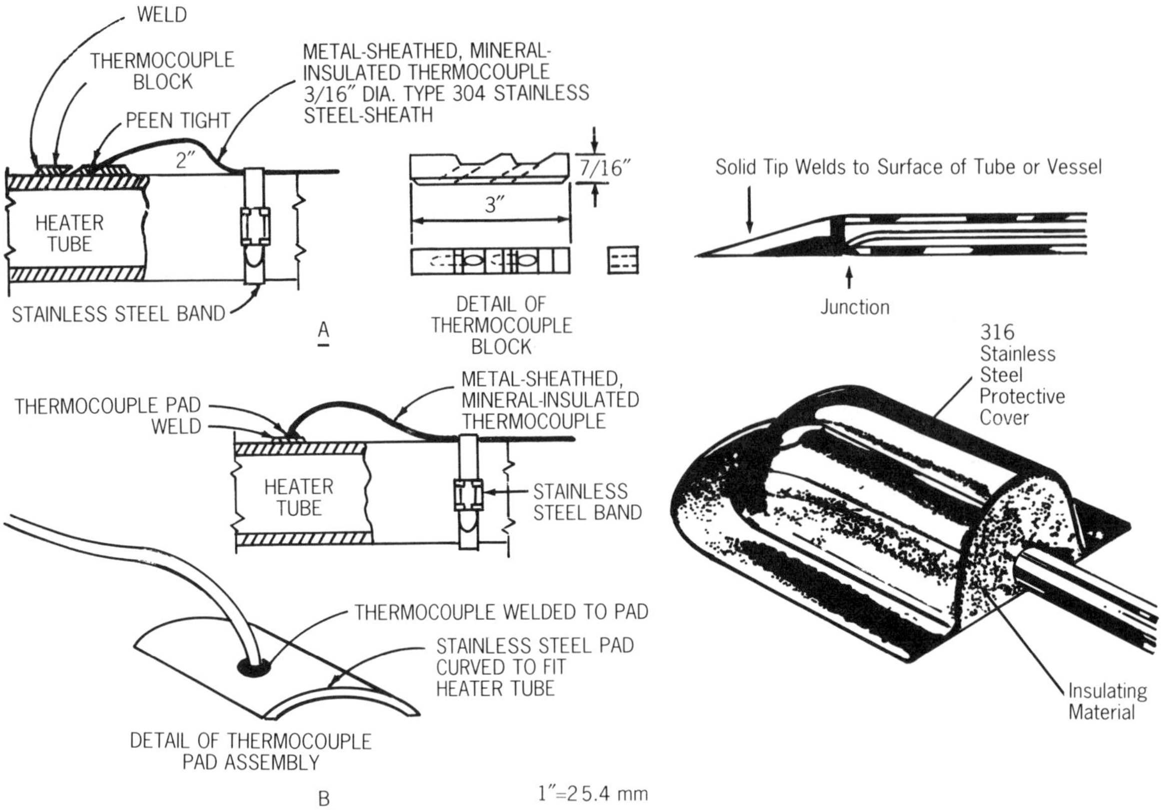

FIG. 16t
Tube surface temperature measured by thermocouple block (A) by welded on stainless steel pad,[2] or by directly welding the thermocouple to the surface (C). A protective cover gives the required mechanical protection.

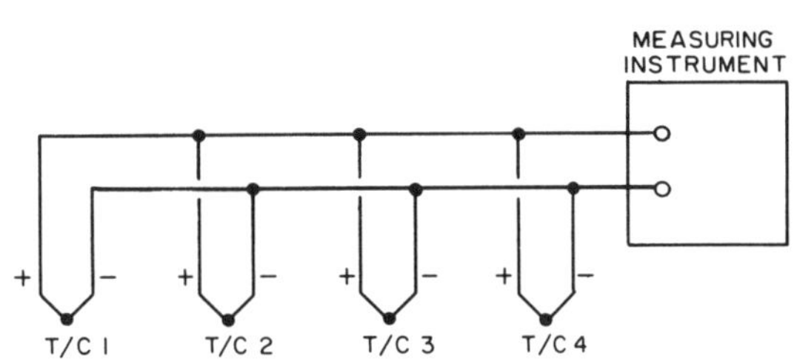

FIG. 16u
Average temperature measurement.

with proper precautions, by two separate measuring instruments (Figure 16w). Generally speaking, a potentiometer draws no current from the thermocouple circuit at balance. With burnout, a small current circulates through the thermocouple. The control point setting also has a bearing. A thermocouple is now available with a built in current transmitter.

Since all thermocouples are subject to drift, calibration checks are done regularly in laboratories and industrial plants.

Thermopiles are thermocouples connected in series with electrically insulated junctions (Figure 1p). Thermopiles generate large emf's, thus reducing sensitivity requirements in the readout instrument. To obtain the mean temperature at several points being monitored by similar thermocouples in

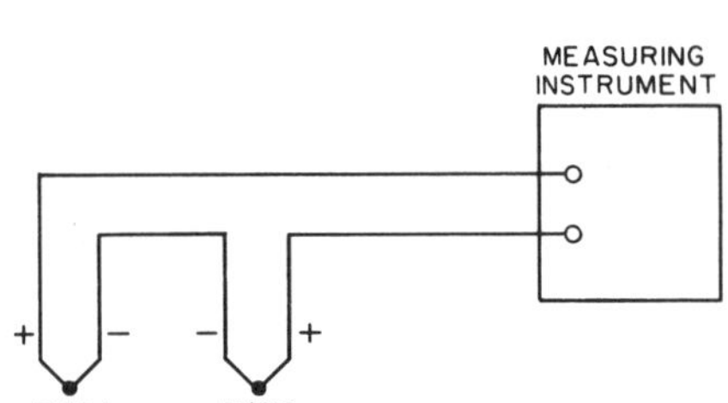

FIG. 16v
Temperature difference.

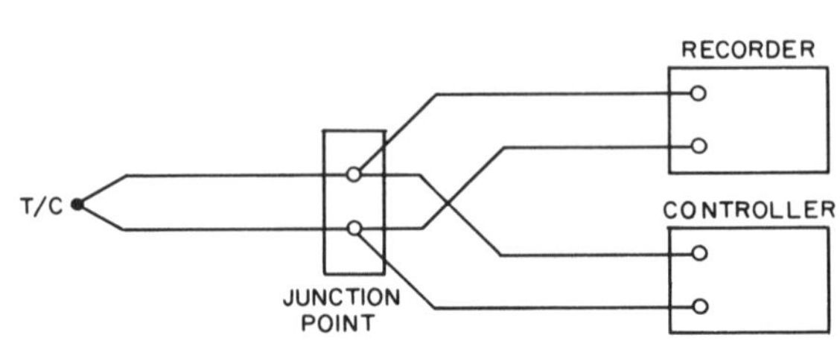

FIG. 16w
Parallel operation from common thermocouple.

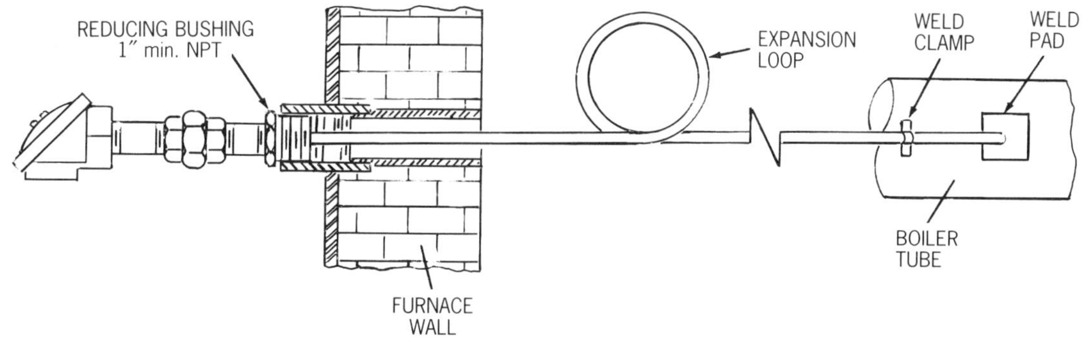

FIG. 16x
The expansion loop allows for thermal expansion in furnace applications.

series, divide the total emf by the number of sensing junctions and relate this emf value to a corresponding temperature reading in the emf-temperature table for the type of thermocouple being used. As was discussed in Section 1, thermopiles can be used to amplify the output signals in differential temperature measurements and to serve as heat-flow detectors.

The principal objections to the use of thermopiles are the need for electrical isolation of individual thermocouples and the error that might go unnoticed when the output of one of the thermocouples is reduced by a short circuit. One satisfactory application for thermopiles is to use them as temperature differential detectors.

THERMOCOUPLE BURNOUT

When a thermocouple detects the temperature in furnaces or superheaters (Figure 16x), the extension wire can pass through flames. On high-temperature services thermocouples are provided with platinum, tantalum, or molybdenum sheath materials over the insulation, which can be magnesium or beryllium oxide. Hot spots like a burning coal seam can eventually burn through the sheath and the insulation of the extension wire. When the wires melt, a short develops. This is called thermocouple burnout. Under these conditions the thermocouple no longer indicates the temperature of the initial hot junction; instead, it measures the temperature at the hot spot. One cannot detect thermocouple burnout by reading the millivolt signal. However, if one measures the thermocouple resistance, that will signal a change as a result of burnout (Figure 16y).

Under normal conditions a running record of the thermocouple resistance will show gradual changes with temperature. When the thermal insulation is beginning to fail and a short is beginning to form, the thermocouple resistance will rise first, and when the hot spot burns through and a short is formed, the thermocouple resistance will drop abruptly. The method used to measure the thermocouple resistance is called offset-compensated ohms measurement. As shown on the top of Figure 16z, normally the millivolts (V_1) are measured across the thermocouple. Then, a current source is connected periodically and the millivolts (V_2) are once more

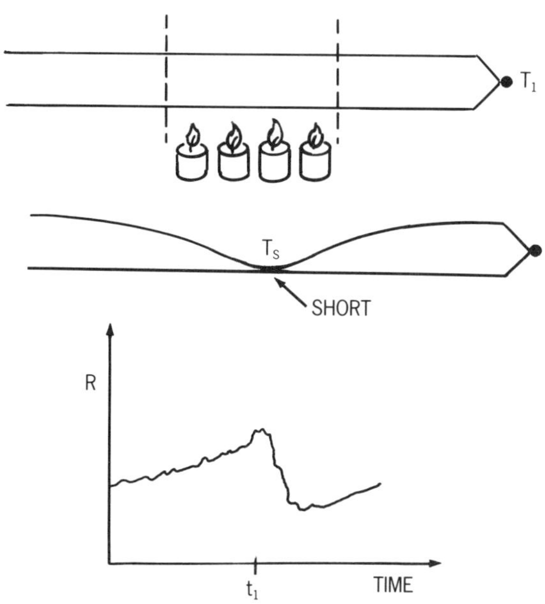

FIG. 16y
Thermocouple burnout can be detected by measuring the resulting drop in thermocouple loop resistance.[1]

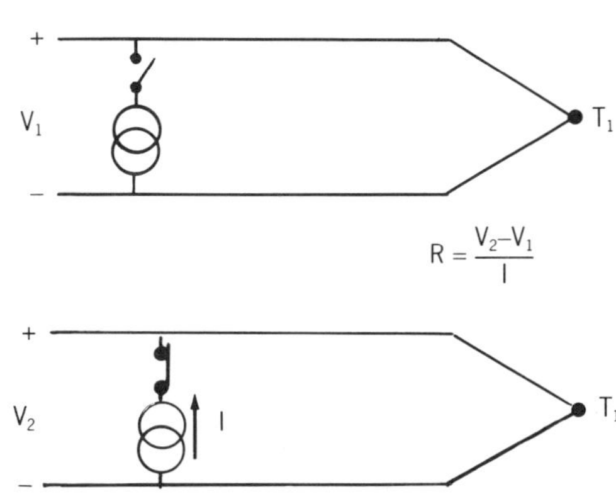

FIG. 16z
Offset-compensated ohms measurement allows detection of the thermocouple loop resistance.

measured. The thermocouple resistance is calculated by subtracting the thermocouple millivolts from the total and dividing it with the current flow in the loop: $R = (V_2 - V_1)/I$. By continuously recording this resistance, one can detect when an abrupt drop occurs, signaling thermocouple burnout.

PROTECTION AGAINST NOISE

The thermocouple signal is very weak—a one degree change in temperature results in only a few millionths of a volt change in output. Because of this, precautions must be taken against errors due to stray currents resulting from the proximity of electrical wiring (common mode noise) or from capacitive secondary grounds (normal mode interference). Common mode noise (Figure 16aa) appears on both thermocouple signal wires and therefore can be filtered out as 60 Hz (or 50 Hz) harmonic noise. The filter does reduce the interference dramatically, but it also causes the voltmeter to be sluggish when responding to a step change. It is also possible to eliminate the common mode interference by using twisted wire leads, because each time the wire is twisted, the flux-induced current is inverted. Another recommended form of protection against any type of common mode noise is guarding and shielding. If the shield surrounding the lead wires is connected to the guard surrounding the voltmeter, the interfering current caused by AC interference does not flow through the thermocouple lead resistance but instead is shunted. Naturally, when thermocouples are scanned, the scanner guard must be switched to the shield of the thermocouple being read to eliminate ground loops. Harmonics can also be removed by integrating the incoming signal over the power line cycle in an integrating analog-to-digital converter or voltmeter. In short, common mode noise is relatively easy to remove.

The same cannot be said about normal mode noise. An example of normal mode noise interferences can occur in the measurement of the temperature in a molten metal bath which is heated by electric current. In this case the thermocouple junction is in direct contact with a common mode noise source, and the capacitive ground (C-stray) from the L_O terminal of the thermocouple to the chassis causes a current flow in the low lead and an associated normal mode noise voltage across the resistance R_s (Figure 16bb). If a guard lead wire is installed connected directly to the thermocouple, the current flowing in the L_O lead through the resistance R_s is drastically reduced. Therefore, the worst form of interference is DC offset caused by a DC leakage current; whatever normal mode noise remains in the system, it cannot be distinguished from the measurement and, in case of weak signals, even a small amount of noise can represent a large amount of interference.

DIAGNOSTICS AND TRANSMISSION

Thermocouple diagnostics can be improved by the use of tip-branched and leg-branched lead wires allowing redundant measurements, verification of system integrity, and other forms of diagnostics. These tools can be useful in

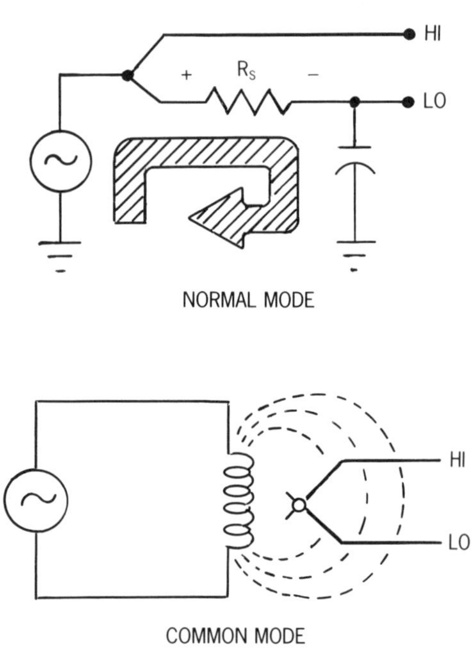

FIG. 16aa
Noise interference that enters only one of the lead wires (normal mode) is more difficult to remove than noise which acts on both leads (common mode).

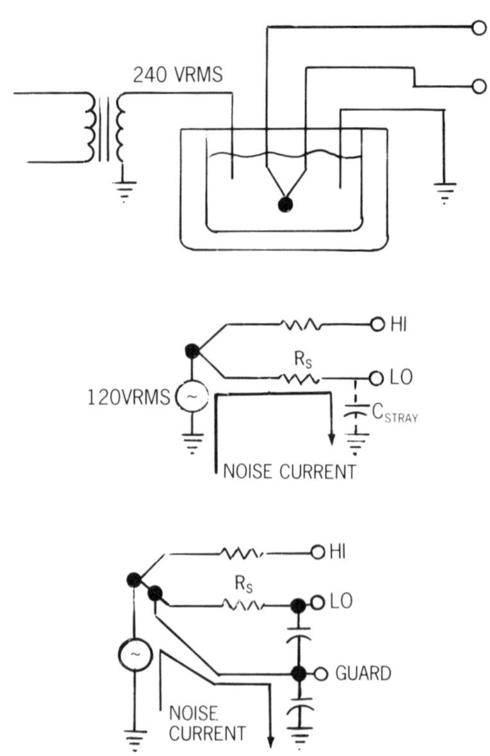

FIG. 16bb
The addition of a guard lead wire reduces the normal mode noise.[1]

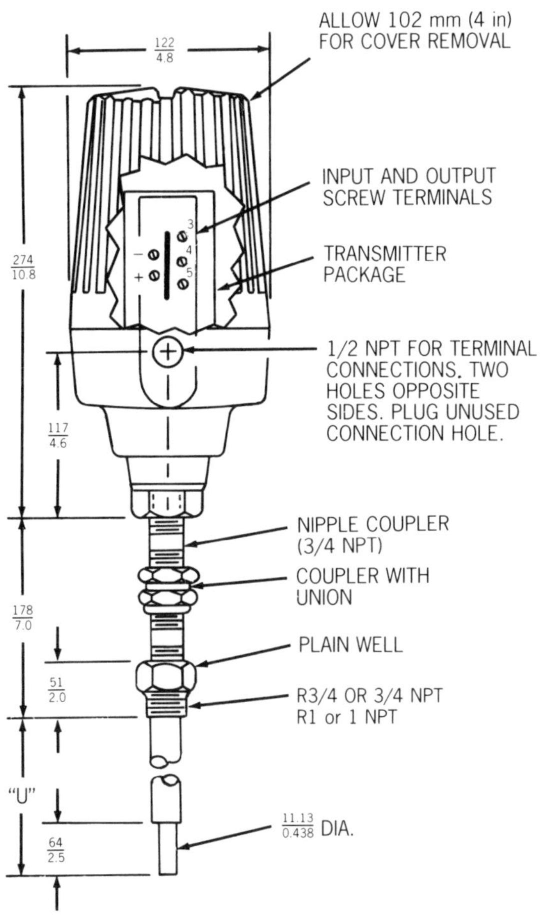

FIG. 16cc
Integral thermocouple transmitter mounted directly on top of the thermowell. (Courtesy of The Foxboro Co.)

detecting the failure of wire insulators, poor junction connections, wire degradation due to overheating, or decalibration due to diffusion of atmospheric particles into the metal. Added to the noise protection and degradation problems are the intermediate wire junctions, which if not at the same temperature (Figure 16e) also contribute errors. Considering all these factors one might conclude that while it is possible to reduce noise interference by guarding, shielding, using twisted pairs, and integration, the best alternative is not to transmit the low-level thermocouple signal over long distances but to place the transmitter electronics directly on top of the thermowell (Figure 16cc).

In the past it was more economical to run the thermocouple lead wires to the data acquisition systems, and this is still the case in the laboratory and on various test stands. On the other hand, the cost of integral transmitters in most industrial applications is competitive with the cost of running the thermocouple lead wires to the control room. These integral transmitters are also explosion-proof. Conventional transmitters are accurate to 0.15% of span, and intelligent units are accurate to 0.05% of span (Table 16m). It can be observed that while intelligent transmitters give better performance than the standard ones, they too are limited to a minimum absolute error plus cold junction error, which equals about 1°F (0.6°C).

In addition to improved performance, the intelligent transmitters are capable of working with any one of eight types of thermocouples or two types of resistance temperature detector (RTD) elements. This increases their flexibility and reduces the need for spare parts. The intelligent transmitters are also provided with continuous self-diagnostics and with automatic three-point self-calibration which is performed every 5 seconds and does not interrupt the analog or digital output of the unit. The intelligent transmitter can also be furnished with dual thermal elements that can be used to measure temperature differentials, averages, and high/low sensors, or as redundant backup elements. Another convenient feature of smart transmitters is their remote reconfiguration capability, which can change their zero, span, or many other features without requiring rewiring.

ADVANTAGES AND LIMITATIONS

Thermocouples are low-cost, small size, rugged, convenient, and versatile (can be welded to a pipe), wide range, reasonably stable, reproducible, accurate, and fast. The emf they generate is independent of wire length and diameter. While RTDs are more accurate and more stable and while thermistors are more sensitive, thermocouples are the most economical sensors and they can detect the highest temperatures.

The main disadvantage of the thermocouple is its weak output signal. This makes it sensitive to electrical noise and limits its use to relatively wide spans (usually the minimum transmitter span is one mV). It is nonlinear, and the conversion of the emf generated into temperature is not as simple as in direct reading devices. Thermocouples always require amplifiers, and their calibration can change due to contamination, composition changes due to internal oxidation, coldworking or temperature gradients. Thermocouples cannot be used bare in conductive fluids, and if their wires are not homogeneous, this can cause emf's that are difficult to detect.

In general, one should use the largest size thermocouple wire possible, avoid stress and vibration, use integral transmitters whenever possible (and otherwise use twisted and shielded wires with the shield connected to the guard of the integrating analog-to-digital converter), avoid steep temperature gradients, and be careful in the selection of sheath and thermowell materials.

TABLE 16dd

Type J—Iron-Constantan Thermocouple

Degrees Fahrenheit vs. Millivolts. Temperatures are based on the International Temperature Scale of 1948. emf is expressed in absolute millivolts. Reference Junction 32°F (0°C). *Note:* Instruments calibrated to this curve have scales identified as "Type J" Thermocouple.

°F*	−300	−200	−100	−0	+0	100	200	300	400	500	600	700	800	900	1000	1100	1200	1300	1400	1500
									MILLIVOLTS											
0	−7.52	−5.76	−3.49	−0.89	−0.89	1.94	4.91	7.94	11.03	14.12	17.18	20.26	23.32	26.40	29.52	32.72	36.01	39.43	42.96	46.53
5	−7.59	−5.86	−3.61	−1.02	−0.75	2.09	5.06	8.10	11.18	14.27	17.34	20.41	23.47	26.55	29.68	32.89	36.18	39.61	43.14	46.71
10	−7.66	−5.96	−3.73	−1.16	−0.61	2.23	5.21	8.25	11.34	14.42	17.49	20.56	23.63	26.70	29.84	33.05	36.35	39.78	43.32	46.89
15	−7.73	−6.06	−3.85	−1.29	−0.48	2.38	5.36	8.40	11.49	14.58	17.64	20.72	23.78	26.86	30.00	33.21	36.52	39.96	43.50	47.07
20	−7.79	−6.16	−3.97	−1.43	−0.34	2.52	5.51	8.56	11.65	14.73	17.80	20.87	23.93	27.02	30.16	33.37	36.69	40.13	43.68	47.24
25		−6.25	−4.09	−1.56	−0.20	2.67	5.66	8.71	11.80	14.88	17.95	21.02	24.09	27.17	30.32	33.54	36.86	40.31	43.85	47.42
30		−6.35	−4.21	−1.70	−0.06	2.82	5.81	8.87	11.96	15.04	18.11	21.18	24.24	27.33	30.48	33.70	37.02	40.48	44.03	47.60
35		−6.44	−4.33	−1.83	+0.08	2.97	5.96	9.02	12.11	15.19	18.26	21.33	24.39	27.48	30.64	33.86	37.20	40.66	44.21	47.78
40		−6.53	−4.44	−1.96	+0.22	3.11	6.11	9.17	12.26	15.34	18.41	21.48	24.55	27.64	30.80	34.03	37.36	40.83	44.39	47.95
45		−6.62	−4.56	−2.09	+0.36	3.26	6.27	9.33	12.42	15.50	18.57	21.64	24.70	27.80	30.96	34.19	37.54	41.01	44.57	48.13
50		−6.71	−4.68	−2.22	0.50	3.41	6.42	9.48	12.57	15.65	18.72	21.79	24.85	27.95	31.12	34.36	37.71	41.19	44.75	48.31
55		−6.80	−4.79	−2.35	0.65	3.56	6.57	9.64	12.73	15.80	18.87	21.94	25.01	28.11	31.28	34.52	37.88	41.36	44.93	48.48
60		−6.89	−4.90	−2.48	0.79	3.71	6.72	9.79	12.88	15.96	19.03	22.10	25.16	28.26	31.44	34.68	38.05	41.54	45.10	48.66
65		−6.97	−5.01	−2.61	0.93	3.86	6.87	9.95	13.04	16.11	19.18	22.25	25.32	28.42	31.60	34.60	38.22	41.72	45.28	48.83
70		−7.06	−5.12	−2.74	1.07	4.01	7.03	10.10	13.19	16.26	19.34	22.40	25.47	28.58	31.76	35.01	38.39	41.90	45.46	49.01
75		−7.14	−5.23	−2.86	1.22	4.16	7.18	10.25	13.34	16.42	19.49	22.55	25.62	28.74	31.92	35.18	38.57	42.07	45.64	49.18
80		−7.22	−5.34	−2.99	1.36	4.31	7.33	10.41	13.50	16.57	19.64	22.71	25.78	28.89	32.08	35.35	38.74	42.25	45.82	49.36
85		−7.30	−5.44	−3.12	1.51	4.46	7.48	10.56	13.65	16.72	19.80	22.86	25.93	29.05	32.24	35.51	38.91	42.43	46.00	49.53
90		−7.38	−5.55	−3.24	1.65	4.61	7.64	10.72	13.81	16.88	19.95	23.01	26.09	29.21	32.40	35.68	39.08	42.61	46.18	49.70
95		−7.45	−5.65	−3.36	1.80	4.76	7.79	10.87	13.96	17.03	20.10	23.17	26.24	29.37	32.56	35.84	39.26	42.78	46.35	49.88
100		−7.52	−5.76	−3.49	1.94	4.91	7.94	11.03	14.12	17.18	20.26	23.32	26.40	29.52	32.72	36.01	39.43	42.96	46.53	50.05

$$*°C = \frac{°F - 32}{1.8}$$

TABLE 16ee

Type K—Chromel-Alumel Thermocouple

Degrees Fahrenheit vs. Millivolts. Temperatures are based on the International Temperature Scale of 1948. emf is expressed in absolute millivolts. Reference Junction 32°F (0°C).

°F*	0	100	200	300	400	500	600	700	800	900	1000	1100	1200	1300	1400	1500	1600	1700	1800	1900	2000	2100	2200	2300
	MILLIVOLTS																							
0	−0.68	1.52	3.82	6.09	8.31	10.57	12.86	15.18	17.53	19.89	22.26	24.63	26.98	29.32	31.65	33.93	36.19	38.43	40.62	42.78	44.91	47.00	49.05	51.05
5	−0.58	1.63	3.94	6.20	8.42	10.68	12.97	15.30	17.64	20.01	22.37	24.74	27.10	29.44	31.76	34.05	36.31	38.54	40.73	42.89	45.01	47.10	49.15	51.15
10	−0.47	1.74	4.05	6.31	8.54	10.79	13.09	15.41	17.76	20.13	22.49	24.86	27.22	29.56	31.88	34.16	36.42	38.65	40.84	42.99	45.12	47.21	49.25	51.25
15	−0.37	1.86	4.17	6.42	8.65	10.91	13.20	15.53	17.88	20.24	22.61	24.98	27.34	29.67	31.99	34.28	36.53	38.76	40.95	43.10	45.22	47.31	49.35	51.35
20	−0.26	1.97	4.28	6.53	8.76	11.02	13.32	15.65	18.00	20.36	22.73	25.10	27.45	29.79	32.11	34.39	36.64	38.87	41.05	43.21	45.33	47.41	49.45	51.45
25	−0.15	2.09	4.40	6.65	8.87	11.13	13.44	15.76	18.11	20.48	22.85	25.22	27.57	29.91	32.22	34.50	36.76	38.98	41.16	43.31	45.43	47.52	49.55	51.54
30	−0.04	2.20	4.51	6.76	8.98	11.25	13.55	15.88	18.23	20.60	22.97	25.34	27.69	30.02	32.34	34.62	36.87	39.09	41.27	43.42	45.54	47.62	49.65	51.64
35	+0.07	2.32	4.63	6.87	9.09	11.36	13.67	16.00	18.35	20.72	23.08	25.46	27.80	30.14	32.45	34.73	36.98	39.20	41.38	43.53	45.64	47.72	49.76	51.74
40	+0.18	2.43	4.74	6.98	9.21	11.48	13.78	16.12	18.47	20.84	23.20	25.57	27.92	30.25	32.57	34.84	37.09	39.31	41.49	43.63	45.75	47.82	49.86	51.84
45	+0.29	2.55	4.86	7.09	9.32	11.59	13.90	16.23	18.58	20.95	23.32	25.69	28.04	30.37	32.68	34.96	37.20	39.42	41.60	43.74	45.85	47.93	49.96	51.94
50	0.40	2.66	4.97	7.20	9.43	11.71	14.02	16.35	18.70	21.07	23.44	25.81	28.15	30.49	32.80	35.07	37.31	39.53	41.70	43.85	45.96	48.03	50.06	52.03
55	0.51	2.78	5.08	7.31	9.54	11.82	14.13	16.47	18.82	21.19	23.56	25.93	28.27	30.60	32.91	35.18	37.43	39.64	41.81	43.95	46.06	48.13	50.16	52.13
60	0.62	2.89	5.20	7.42	9.66	11.94	14.25	16.59	18.94	21.31	23.68	26.05	28.39	30.72	33.02	35.29	37.54	39.75	41.92	44.06	46.17	48.23	50.26	52.23
65	0.73	3.01	5.31	7.53	9.77	12.05	14.36	16.70	19.06	21.43	23.80	26.16	28.50	30.83	33.14	35.41	37.65	39.86	42.03	44.17	46.27	48.34	50.36	52.33
70	0.84	3.12	5.42	7.64	9.88	12.17	14.48	16.82	19.18	21.54	23.91	26.28	28.62	30.95	33.25	35.52	37.76	39.96	42.14	44.27	46.38	48.44	50.46	52.42
75	0.95	3.24	5.53	7.75	10.00	12.28	14.60	16.94	19.29	21.66	24.03	26.40	28.74	31.07	33.37	35.63	37.87	40.07	42.24	44.38	46.48	48.54	50.56	52.52
80	1.06	3.36	5.65	7.87	10.11	12.40	14.71	17.06	19.41	21.78	24.15	26.52	28.86	31.18	33.48	35.75	37.98	40.18	42.35	44.49	46.58	48.64	50.65	52.62
85	1.18	3.47	5.76	7.98	10.22	12.51	14.83	17.17	19.53	21.90	24.27	26.63	28.97	31.30	33.59	35.86	38.09	40.29	42.46	44.59	46.69	48.74	50.75	52.72
90	1.29	3.59	5.87	8.09	10.34	12.63	14.95	17.29	19.65	22.02	24.39	26.75	29.09	31.42	33.71	35.97	38.20	40.40	42.57	44.70	46.79	48.85	50.85	52.81
95	1.40	3.70	5.98	8.20	10.45	12.74	15.06	17.41	19.77	22.14	24.51	26.87	29.21	31.53	33.82	36.08	38.32	40.51	42.67	44.80	46.90	48.95	50.95	52.91
100	1.52	3.82	6.09	8.31	10.57	12.86	15.18	17.53	19.89	22.26	24.63	26.98	29.32	31.65	33.93	36.19	38.43	40.62	42.78	44.91	47.00	49.05	51.05	53.01

$*°C = \frac{°F - 32}{1.8}$

TABLE 16ff

Type R—Platinum vs. Platinum Plus 13% Rhodium Thermocouple

Degrees Fahrenheit vs. Millivolts. Temperatures are based on the International Temperature Scale of 1948. emf is expressed in absolute millivolts. Reference Junction 32°F (0°C).

°F*	0	100	200	300	400	500	600	700	800	900	1000	1100	1200	1300	1400	1500
								MILLIVOLTS								
0	−0.089	0.220	0.596	1.030	1.504	2.012	2.547	3.103	3.677	4.264	4.868	5.488	6.125	6.773	7.436	8.116
5	−0.076	0.237	0.616	1.052	1.529	2.038	2.575	3.132	3.706	4.294	4.899	5.519	6.156	6.805	7.470	8.150
10	−0.062	0.255	0.637	1.075	1.553	2.065	2.602	3.160	3.735	4.324	4.930	5.551	6.188	6.838	7.503	8.184
15	−0.049	0.272	0.657	1.098	1.578	2.091	2.630	3.188	3.764	4.354	4.960	5.582	6.220	6.871	7.537	8.218
20	−0.035	0.291	0.678	1.121	1.603	2.117	2.657	3.217	3.794	4.384	4.991	5.614	6.252	6.904	7.571	8.253
25	−0.021	0.308	0.700	1.144	1.628	2.144	2.685	3.245	3.823	4.413	5.022	5.645	6.285	6.937	7.605	8.287
30	−0.006	0.327	0.721	1.167	1.653	2.170	2.712	3.273	3.852	4.443	5.053	5.677	6.317	6.970	7.639	8.322
35	+0.009	0.345	0.742	1.191	1.678	2.197	2.740	3.302	3.882	4.473	5.084	5.709	6.349	7.003	7.672	8.356
40	+0.024	0.363	0.763	1.214	1.703	2.223	2.768	3.330	3.911	4.503	5.115	5.741	6.381	7.037	7.706	8.391
45	+0.039	0.381	0.785	1.238	1.729	2.250	2.796	3.359	3.941	4.533	5.146	5.773	6.414	7.069	7.740	8.426
50	0.055	0.400	0.807	1.261	1.754	2.277	2.823	3.387	3.970	4.563	5.176	5.805	6.446	7.103	7.774	8.460
55	0.071	0.419	0.828	1.285	1.779	2.303	2.851	3.416	3.999	4.593	5.208	5.837	6.479	7.136	7.808	8.495
60	0.086	0.438	0.850	1.309	1.805	2.330	2.879	3.445	4.029	4.624	5.238	5.869	6.511	7.169	7.842	8.530
65	0.103	0.457	0.872	1.333	1.831	2.357	2.907	3.473	4.058	4.654	5.270	5.901	6.544	7.202	7.877	8.565
70	0.119	0.476	0.894	1.357	1.856	2.384	2.935	3.502	4.087	4.685	5.301	5.933	6.577	7.235	7.911	8.599
75	0.135	0.496	0.917	1.381	1.882	2.412	2.963	3.531	4.116	4.715	5.332	5.964	6.609	7.269	7.945	8.634
80	0.152	0.516	0.939	1.406	1.908	2.438	2.991	3.560	4.146	4.746	5.363	5.996	6.642	7.302	7.979	8.669
85	0.169	0.536	0.962	1.430	1.934	2.466	3.019	3.589	4.175	4.776	5.394	6.028	6.674	7.336	8.013	8.704
90	0.186	0.556	0.984	1.455	1.960	2.493	3.047	3.618	4.205	4.807	5.426	6.060	6.707	7.369	8.047	8.739
95	0.203	0.576	1.007	1.480	1.986	2.520	3.075	3.647	4.235	4.837	5.457	6.092	6.740	7.403	8.081	8.774
100	0.220	0.596	1.030	1.504	2.012	2.547	3.103	3.677	4.264	4.868	5.488	6.125	6.773	7.436	8.116	8.809

$*°C = \frac{°F - 32}{1.8}$

TABLE 16ff Continued

Type R—Platinum vs. Platinum Plus 13% Rhodium Thermocouple

°F*	1600	1700	1800	1900	2000	2100	2200	2300	2400	2500	2600	2700	2800	2900	3000
							MILLIVOLTS								
0	8.809	9.516	10.237	10.973	11.726	12.488	13.255	14.027	14.798	15.568	16.340	17.110	17.875	18.636	19.394
5	8.844	9.552	10.274	11.011	11.765	12.526	13.293	14.065	14.837	15.607	16.378	17.148	17.913	18.674	19.432
10	8.879	9.587	10.310	11.048	11.802	12.564	13.332	14.104	14.875	15.645	16.417	17.186	17.951	18.712	19.470
15	8.914	9.623	10.347	11.085	11.840	12.602	13.371	14.142	14.914	15.684	16.455	17.225	17.989	18.750	19.508
20	8.949	9.659	10.383	11.122	11.878	12.641	13.409	14.181	14.952	15.722	16.494	17.263	18.027	18.788	19.545
25	8.984	9.694	10.420	11.160	11.916	12.679	13.448	14.219	14.991	15.761	16.532	17.301	18.065	18.826	19.583
30	9.019	9.730	10.456	11.197	11.954	12.718	13.486	14.258	15.029	15.800	16.571	17.340	18.103	18.864	19.621
35	9.054	9.766	10.493	11.235	11.992	12.756	13.525	14.296	15.068	15.838	16.610	17.378	18.141	18.902	19.659
40	9.090	9.802	10.529	11.273	12.029	12.795	13.564	14.335	15.107	15.877	16.648	17.416	18.179	18.940	19.697
45	9.125	9.838	10.566	11.310	12.068	12.833	13.602	14.374	15.145	15.915	16.687	17.455	18.218	18.978	19.735
50	9.161	9.874	10.603	11.348	12.105	12.871	13.641	14.412	15.184	15.954	16.725	17.493	18.255	19.016	19.773
55	9.196	9.910	10.639	11.385	12.144	12.909	13.679	14.451	15.222	15.992	16.764	17.532	18.294	19.054	19.811
60	9.232	9.946	10.676	11.424	12.182	12.948	13.718	14.490	15.261	16.031	16.802	17.569	18.332	19.092	19.848
65	9.267	9.982	10.712	11.461	12.220	12.986	13.756	14.528	15.299	16.070	16.842	17.608	18.370	19.129	19.886
70	9.303	10.019	10.749	11.499	12.258	13.025	13.795	14.567	15.338	16.108	16.880	17.646	18.408	19.168	19.924
75	9.338	10.056	10.786	11.537	12.296	13.063	13.833	14.606	15.377	16.147	16.918	17.685	18.446	19.205	19.962
80	9.374	10.092	10.823	11.575	12.335	13.102	13.872	14.644	15.415	16.185	16.957	17.723	18.484	19.243	19.999
85	9.409	10.129	10.861	11.613	12.373	13.140	13.911	14.683	15.454	16.224	16.995	17.761	18.522	19.281	20.037
90	9.445	10.164	10.898	11.651	12.411	13.178	13.949	14.721	15.492	16.263	17.033	17.799	18.560	19.318	20.075
95	9.481	10.201	10.936	11.689	12.450	13.216	13.988	14.760	15.531	16.301	17.072	17.837	18.598	19.356	20.112
100	9.516	10.237	10.973	11.726	12.488	13.255	14.027	14.798	15.568	16.340	17.110	17.875	18.636	19.394	20.150

$*°C = \frac{°F - 32}{1.8}$

TABLE 16gg

Type S—Platinum vs. Platinum Plus 10% Rhodium Thermocouple

Degrees Fahrenheit vs. Millivolts. Temperatures are based on the International Temperature Scale of 1948. emf is expressed in absolute millivolts. Reference Junction 32°F (0°C).

°F*	0	100	200	300	400	500	600	700	800	900	1000	1100	1200	1300	1400	1500
								MILLIVOLTS								
0	−0.092	0.221	0.595	1.017	1.474	1.956	2.458	2.977	3.506	4.046	4.596	5.156	5.726	6.307	6.897	7.498
5	−0.078	0.238	0.615	1.039	1.498	1.981	2.484	3.003	3.533	4.073	4.623	5.184	5.755	6.336	6.927	7.529
10	−0.064	0.256	0.635	1.061	1.521	2.005	2.510	3.029	3.560	4.100	4.651	5.212	5.784	6.635	6.957	7.559
15	−0.050	0.274	0.655	1.083	1.545	2.030	2.535	3.056	3.587	4.128	4.679	5.241	5.813	6.394	6.987	7.589
20	−0.035	0.291	0.676	1.106	1.569	2.055	2.561	3.082	3.614	4.155	4.707	5.269	5.842	6.424	7.017	7.620
25	−0.021	0.309	0.696	1.128	1.593	2.080	2.587	3.108	3.640	4.182	4.735	5.298	5.871	6.453	7.046	7.650
30	−0.006	0.327	0.717	1.151	1.616	2.105	2.613	3.135	3.667	4.210	4.763	5.326	5.899	6.483	7.076	7.681
35	+0.009	0.346	0.738	1.173	1.640	2.130	2.638	3.161	3.694	4.237	4.790	5.354	5.928	6.512	7.106	7.711
40	+0.024	0.364	0.758	1.196	1.664	2.155	2.664	3.188	3.721	4.264	4.818	5.383	5.957	6.542	7.136	7.742
45	+0.040	0.383	0.779	1.219	1.688	2.180	2.690	3.214	3.748	4.292	4.846	5.411	5.986	6.571	7.166	7.772
50	0.056	0.401	0.800	1.242	1.712	2.205	2.716	3.240	3.775	4.319	4.874	5.440	6.015	6.601	7.196	7.803
55	0.071	0.420	0.822	1.264	1.736	2.230	2.742	3.267	3.802	4.347	4.902	5.469	6.044	6.630	7.226	7.834
60	0.087	0.439	0.843	1.287	1.761	2.255	2.768	3.293	3.829	4.374	4.930	5.497	6.073	6.660	7.257	7.864
65	0.104	0.458	0.864	1.311	1.785	2.281	2.794	3.320	3.856	4.402	4.959	5.526	6.102	6.689	7.287	7.895
70	0.120	0.477	0.886	1.334	1.809	2.306	2.820	3.347	3.883	4.430	4.987	5.555	6.131	6.719	7.317	7.925
75	0.136	0.496	0.907	1.357	1.833	2.331	2.846	3.373	3.910	4.457	5.015	5.583	6.161	6.749	7.347	7.956
80	0.153	0.516	0.929	1.380	1.858	2.357	2.872	3.400	3.937	4.485	5.043	5.612	6.190	6.778	7.377	7.987
85	0.170	0.535	0.951	1.404	1.882	2.382	2.898	3.426	3.964	4.512	5.071	5.640	6.219	6.808	7.407	8.018
90	0.187	0.555	0.973	1.427	1.907	2.407	2.924	3.453	3.991	4.540	5.099	5.669	6.248	6.838	7.438	8.048
95	0.204	0.575	0.994	1.450	1.931	2.433	2.951	3.480	4.019	4.568	5.128	5.698	6.277	6.867	7.468	8.079
100	0.221	0.595	1.017	1.474	1.956	2.458	2.977	3.506	4.046	4.596	5.156	5.726	6.307	6.897	7.498	8.110

$*{}^\circ C = \frac{{}^\circ F - 32}{1.8}$

TABLE 16gg Continued

Type S—Platinum vs. Platinum Plus 10% Rhodium Thermocouple

°F*	1600	1700	1800	1900	2000	2100	2200	2300	2400	2500	2600	2700	2800	2900	3000	3100	3200
								MILLIVOLTS									
0	8.110	8.732	9.365	10.009	10.662	11.323	11.989	12.657	13.325	13.991	14.656	15.319	15.979	16.637	17.292	17.943	18.590
5	8.141	8.764	9.397	10.041	10.695	11.356	12.022	12.690	13.358	14.024	14.689	15.352	16.012	16.670	17.324	17.975	18.622
10	8.172	8.795	9.429	10.074	10.728	11.389	12.055	12.724	13.391	14.058	14.722	15.385	16.045	16.702	17.357	18.008	18.655
15	8.203	8.827	9.461	10.106	10.761	11.423	12.089	12.757	13.425	14.091	14.755	15.418	16.078	16.735	17.389	18.040	18.687
20	8.234	8.858	9.493	10.139	10.794	11.456	12.122	12.790	13.458	14.124	14.789	15.451	16.111	16.768	17.422	18.073	
25	8.265	8.890	9.525	10.171	10.827	11.489	12.155	12.824	13.491	14.157	14.822	15.484	16.144	16.801	17.455	18.105	
30	8.296	8.921	9.557	10.204	10.860	11.522	12.189	12.857	13.525	14.191	14.855	15.517	16.177	16.834	17.487	18.137	
35	8.327	8.953	9.589	10.237	10.893	11.556	12.222	12.891	13.558	14.224	14.888	15.550	16.210	16.866	17.520	18.170	
40	8.358	8.984	9.621	10.269	10.926	11.589	12.256	12.924	13.591	14.257	14.921	15.583	16.243	16.899	17.552	18.202	
45	8.389	9.016	9.654	10.302	10.959	11.622	12.289	12.957	13.625	14.290	14.954	15.616	16.275	16.932	17.585	18.235	
50	8.420	9.048	9.686	10.334	10.992	11.655	12.322	12.991	13.658	14.324	14.988	15.649	16.308	16.965	17.618	18.267	
55	8.451	9.079	9.718	10.367	11.025	11.689	12.356	13.024	13.691	14.357	15.021	15.682	16.341	16.997	17.650	18.299	
60	8.482	9.111	9.750	10.400	11.058	11.722	12.389	13.058	13.725	14.390	15.054	15.715	16.374	17.030	17.683	18.332	
65	8.513	9.143	9.782	10.433	11.091	11.755	12.423	13.091	13.758	14.423	15.087	15.748	16.407	17.063	17.715	18.364	
70	8.545	9.174	9.815	10.465	11.124	11.789	12.456	13.124	13.791	14.457	15.120	15.781	16.440	17.095	17.748	18.396	
75	8.576	9.206	9.847	10.498	11.157	11.822	12.490	13.158	13.825	14.490	15.153	15.814	16.473	17.128	17.780	18.429	
80	8.607	9.238	9.879	10.531	11.190	11.855	12.523	13.191	13.858	14.523	15.186	15.847	16.506	17.161	17.813	18.461	
85	8.638	9.270	9.912	10.564	11.224	11.888	12.556	13.224	13.891	14.556	15.219	15.880	16.538	17.194	17.845	18.493	
90	8.670	9.302	9.944	10.597	11.257	11.922	12.590	13.258	13.924	14.589	15.253	15.913	16.571	17.226	17.878	18.526	
95	8.701	9.333	9.976	10.629	11.290	11.955	12.623	13.291	13.958	14.623	15.286	15.946	16.604	17.259	17.910	18.558	
100	8.732	9.365	10.009	10.662	11.323	11.989	12.657	13.325	13.991	14.656	15.319	15.979	16.637	17.292	17.943	18.590	

$*°C = \frac{°F - 32}{1.8}$

TABLE 16hh

Type T—Copper-Constantan Thermocouple

Degrees Fahrenheit vs. Millivolts. Temperatures are based on the International Temperature Scale of 1948. emf is expressed in absolute millivolts. Reference Junction 32°F (0°C).

°F*	−300	−200	−100	−0	+0	100	200	300	400	500	600	700
					MILLIVOLTS							
0	−5.284	−4.111	−2.559	−0.670	−0.670	1.517	3.967	6.647	9.525	12.575	15.773	19.100
5	−5.332	−4.179	−2.645	−0.771	−0.567	1.633	4.096	6.786	9.674	12.732	15.937	19.269
10	−5.379	−4.246	−2.730	−0.872	−0.463	1.751	4.225	6.926	9.823	12.888	16.101	19.439
15		−4.312	−2.814	−0.973	−0.359	1.869	4.355	7.066	9.973	13.046	16.264	19.608
20		−4.377	−2.897	−1.072	−0.254	1.987	4.486	7.208	10.123	13.203	16.429	19.779
25		−4.441	−2.980	−1.171	−0.149	2.107	4.617	7.349	10.273	13.362	16.593	19.949
30		−4.504	−3.062	−1.270	−0.042	2.226	4.749	7.491	10.423	13.520	16.758	20.120
35		−4.566	−3.143	−1.367	+0.064	2.346	4.880	7.633	10.574	13.678	16.924	20.291
40		−4.627	−3.223	−1.463	+0.171	2.467	5.014	7.776	10.726	13.838	17.089	20.463
45		−4.688	−3.301	−1.559	+0.280	2.589	5.147	7.920	10.878	13.997	17.255	20.634
50		−4.747	−3.380	−1.654	0.389	2.711	5.280	8.064	11.030	14.157	17.421	20.805
55		−4.805	−3.457	−1.748	0.499	2.835	5.415	8.207	11.183	14.317	17.588	
60		−4.863	−3.533	−1.842	0.609	2.958	5.550	8.352	11.336	14.477	17.754	
65		−4.919	−3.609	−1.934	0.720	3.082	5.685	8.497	11.490	14.637	17.921	
70		−4.974	−3.684	−2.026	0.832	3.207	5.821	8.642	11.643	14.799	18.089	
75		−5.029	−3.757	−2.117	0.944	3.332	5.957	8.788	11.797	14.961	18.257	
80		−5.081	−3.829	−2.207	1.057	3.458	6.094	8.935	11.953	15.122	18.425	
85		−5.134	−3.901	−2.296	1.171	3.584	6.232	9.082	12.108	15.284	18.593	
90		−5.185	−3.972	−2.385	1.286	3.712	6.370	9.229	12.263	15.447	18.761	
95		−5.235	−4.042	−2.472	1.401	3.839	6.508	9.376	12.418	15.610	18.930	
100		−5.284	−4.111	−2.559	1.517	3.967	6.647	9.525	12.575	15.773	19.100	

$*°C = \frac{°F - 32}{1.8}$

TABLE 16ii

Type E—Chromel-Constantan Thermocouple

°F*	−300	−200	−100	−0	+0	100	200	300	400	500	600	700
					MILLIVOLTS							
0	−8.30	−6.40	−3.94	−1.02	−1.02	2.27	5.87	9.71	13.75	17.95	22.25	26.65
10	−8.45	−6.62	−4.21	−1.33	−0.71	2.62	6.25	10.11	14.17	18.38	22.69	27.09
20	−8.60	−6.83	−4.47	−1.64	−0.39	2.97	6.62	10.51	14.59	18.81	23.13	27.53
30		−7.04	−4.73	−1.94	−0.07	3.32	7.00	10.91	15.00	19.23	23.57	27.97
40		−7.24	−4.98	−2.24	0.26	3.68	7.38	11.31	15.42	19.66	24.00	28.42
50		−7.44	−5.23	−2.54	0.59	4.04	7.76	11.71	15.84	20.09	24.44	28.86
60		−7.62	−5.48	−2.83	0.92	4.40	8.15	12.11	16.26	20.52	24.88	29.31
70		−7.80	−5.72	−3.11	1.26	4.77	8.54	12.52	16.68	20.95	25.32	29.75
80		−7.97	−5.95	−3.39	1.59	5.13	8.93	12.93	17.10	21.39	25.76	30.19
90		−8.14	−6.18	−3.67	1.93	5.50	9.32	13.34	17.52	21.82	26.20	30.64
100		−8.30	−6.40	−3.94	2.27	5.87	9.71	13.75	17.95	22.25	26.65	31.09

°F*	800	900	1000	1100	1200	1300	1400	1500	1600	1700	1800
						MILLIVOLTS					
0	31.09	35.57	40.06	44.56	49.04	53.50	57.92	62.30	66.63	70.90	75.12
10	31.54	36.02	40.51	45.01	49.49	53.94	58.36	62.74	67.05	71.32	75.53
20	31.98	36.47	40.96	45.46	49.93	54.38	58.80	63.17	67.48	71.75	75.95
30	32.43	36.92	41.41	45.91	50.37	54.83	59.24	63.60	67.91	72.17	76.37
40	32.87	37.37	41.86	46.36	50.82	55.27	59.68	64.04	68.34	72.60	
50	33.32	37.82	42.31	46.81	51.27	55.71	60.11	64.47	68.76	73.02	
60	33.77	38.26	42.76	47.26	51.72	56.15	60.55	64.90	69.19	73.44	
70	34.22	38.71	43.21	47.71	52.16	56.59	60.99	65.34	69.62	73.86	
80	34.67	39.16	43.66	48.15	52.61	57.03	61.43	65.77	70.05	74.28	
90	35.12	39.61	44.11	48.60	53.05	57.48	61.86	66.20	70.47	74.70	
100	35.57	40.06	44.56	49.04	53.50	57.92	62.30	66.63	70.90	75.12	

$*°C = \frac{°F - 32}{1.8}$

References

1. ''Practical Temperature Measurement,'' Application Note 290, Hewlett-Packard.
2. ''Installation of Refinery Instruments and Control Systems,'' American Petroleum Institute, Manual API 550.

Bibliography

''American National Standard for Temperature Measurement Thermocouples,'' ANSI MC 96.7, 1975.

Anderson, R.L., ''Accuracy of Small Diameter Sheathed Thermocouples for the Core Flow Test Loop,'' Oak Ridge National Laboratories, ORNL-5401, April 1979.

Ball, K., ''Thermocouples and RTD's,'' *InTech,* August 1986.

Bartosiak, G., ''Guide to Thermocouples,'' *Instruments & Control Systems,* November 1978, pp. 53–55.

Bediones, D., and Wang, T.P., ''Criteria for the Selection of Thermocouples Versus RTD's in Industrial Applications,'' 1991 ISA Conference, Paper #91-0300.

Behrmann, W.C., ''Thermocouple Errors Due To Sheath Conduction,'' *InTech,* August 1990.

Carlson, D.R., ''Temperature Measurement in Process Control,'' *InTech,* October 1990.

Carlson, D.R., ''Transmitting a Remote Temperature,'' *Control,* April 1989.

Coffee, M.B., ''Common-Mode Rejection Techniques for Low-Level Data Acquisition,'' *Instrumentation Technology,* July 1977, pp. 45–49.

Daneman, H.L., ''MIMS Thermocouple Stability,'' *Measurements and Control,* February 1991.

Kaufman, A.B., ''Temperature Calibration,'' *Measurements and Control,* February 1986.

Klipec, B., ''How to Avoid Noise Pickup on Wire and Cable,'' *Instruments & Control Systems,* December 1977, pp. 27–30.

Kreider, K., ''Thermocouple Research at NIST,'' 1991 ISA Conference, Paper #91-0301.

Leewis, W., ''The International Temperature Scale of 1990,'' 1991 ISA Conference, Paper #91-0302.

''Manual on the Use of Thermocouples in Temperature Measurement,'' ASTM Special Publication 470A, 1974.

Moffat, R.J., ''The Gradient Approach to Thermocouple Circuitry,'' in *Temperature, Its Measurement and Control in Science and Industry,* New York: Van Nostrand Reinhold, 1962.

Park, R.M., ''Applying the Systems Concept to Thermocouples,'' *Instrumentation Technology,* August 1973, pp. 25–31.

Pederson, R., ''Choosing Thermocouples or Resistance Thermometers,'' *Instruments & Control Systems,* June 1975, p. 39.

Peterson, W., ''Choosing the Right Temperature Transmitter,'' *InTech,* April 1991.

Reed, R.P., ''Branched Thermocouple Circuits in Underground Coal Gasification Experiments,'' 22nd ISA Symposium, 1976.

''Thermocouples,'' *Measurements and Control,* June 1991.

Wang, T.P., ''Thermocouple Selection and Maintenance,'' *InTech,* April 1991.

Warmoth, D., ''Temperature Measurement & Control,'' *Measurements and Control,* February 1986.

Wither, P., ''Thermocouple Calibration,'' *Measurements and Control,* September 1990.

17 Thermowells

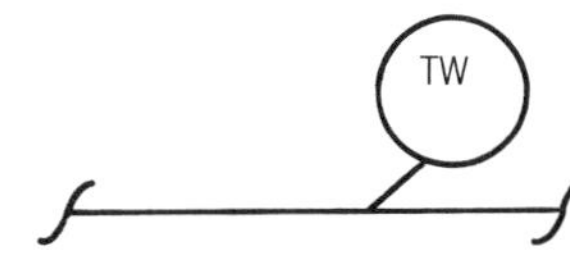

Flow Sheet Symbol

H. L. DANEMAN (1993)

Applications:	Mechanical and corrosion protection of thermal elements and sighting of radiation pyrometers
Temperature Range:	From cryogenic up to 3500°F (1950°C)
Materials of Construction:	Carbon steel, 304 or 316 stainless steel, Carpenter 20, Hastelloy B or C, Inconel, Monel, nickel, titanium, glass-lined, or ceramics (alumina or mullite)
Costs:	Steel, 304 or 316 stainless steel well with 0.5 in. (13 mm) NPT threaded connection and 2.5 in. (64 mm) insertion length costs \$30, while with 10.5 in. (268 mm) insertion length the cost is about \$70. Flanged units cost about twice as much. Relative to stainless steel, the cost multipliers of other well materials are as follows: 1.1 for Monel; 2.0 for Inconel, Carpenter 20, nickel, titanium, and ceramics; and 3.0 for Hastelloy B and C
Partial List of Suppliers:	Advanced Dynamics; Alloy Engineering Co. Inc.; Barber Colman Co.; Chino Works America Inc.; Conax Buffalo Corp.; Degussa Corp.; Devar Inc.; Dresser Industrial Instrument Div.; The Foxboro Co.; Gay Engineering and Sales; Helicoid Div. of Bristol Babcock Inc.; Heraeus Sensor; Honeywell Industrial Controls; Leeds & Northrup, a Unit of General Signal; Moeller Instrument Co. Inc.; Noral Inc.; Omega Engineering; Pyrocom Co.; Sandelius Instruments Inc.; Siemens Energy & Automation; STI Manufacturing Inc.; Tel-Tru Manufacturing Co.; Temp-Pro Inc.; Thermo-Kinetics Co. Ltd.; Thermo Electric (Canada) Ltd.; Trend Instruments Inc.; Tudor Technology Inc.; Valmet Automation Inc.

The purpose of thermowells is to protect the thermal elements from mechanical damage and corrosion or to act as sighting tubes for radiation thermometers. Some of these protective sockets have already been discussed in other sections of this volume; namely, the protection of bimetallic elements was covered in connection with Figure 2c, filled bulbs on Figure 6e, pyrometer sighting tubes on Figure 11t, RTD protection on Figures 13j and 13l, and thermocouple wells on Figures 16q and 16s. The discussion here will build on (but not repeat) what was already said in the noted sections.

Thermowells permit elements to be removed for calibration, replacement or repairs, or the use of portable sensors. The well or protection tube is fixed into the pipe or vessel secured by threads, flanges, or welds allowing the element to be inserted without stress (Figure 17a).

THERMOWELL TYPES

Thermowells are usually metallic and may be coated with other materials to provide additional corrosion protection. Integral flanges or threaded sections enable the wells to be

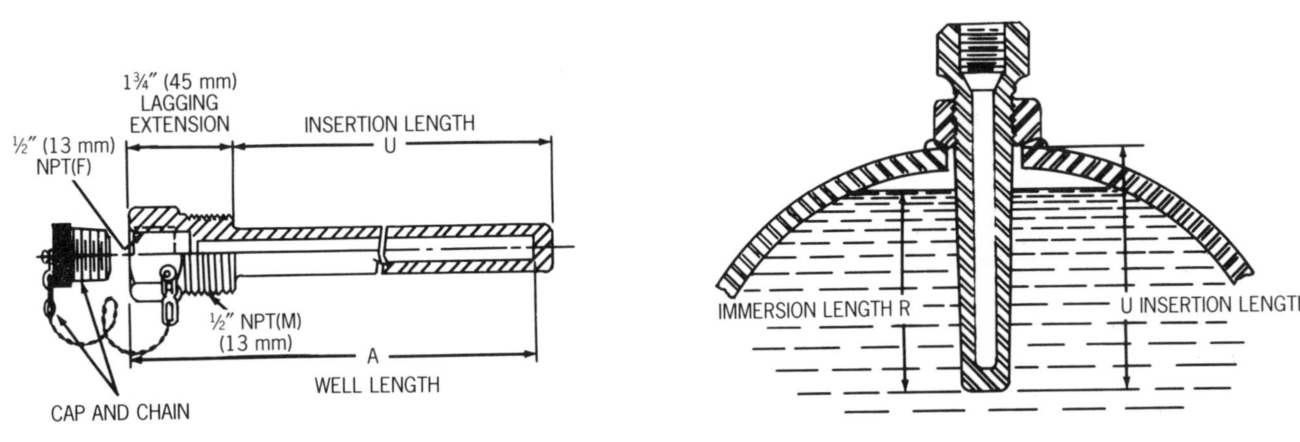

FIG. 17a
The test well and its installation.

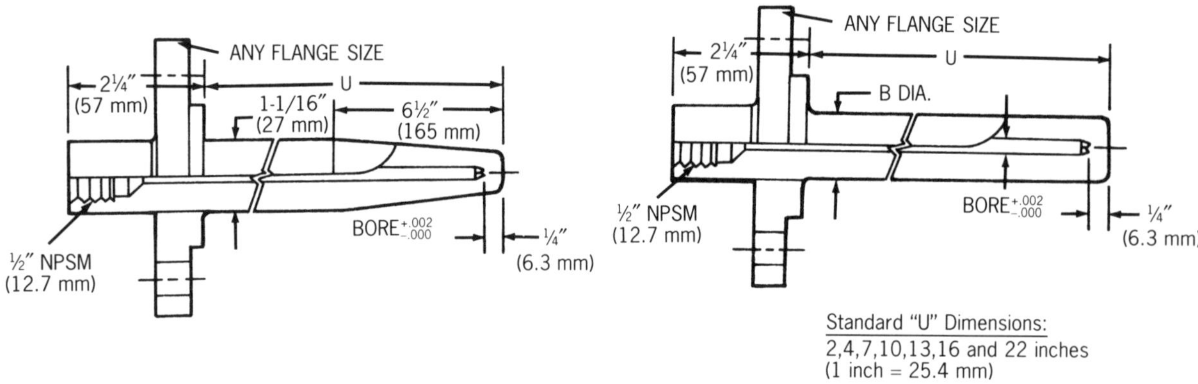

FIG. 17b
Flanged thermowells with tapered and straight shanks. (Courtesy of Trend Instruments Inc.)

secured to the pipe or vessel (Figure 17b). Internal threads secure nipples which extend the head beyond any insulation. It is important to avoid contamination of temperature sensors by oil (such as cutting oils) left on the inner surface of drilled wells or accumulations of foreign materials within wells.

Protection tubes can also be ceramic when used to protect noble metal thermocouples or as sighting tubes for radiation pyrometers. While not as strong as metallic thermowells, they do not droop and can withstand higher temperatures; in addition, they are free of contamination which can cause thermocouple drift due to vapor deposition of elements at higher temperatures. Mullite and high-purity alumina are commonly used as ceramic thermowell materials. In addition to being corrosion-resistant, mullite can operate up to 3200°F (1750°C) and alumina up to 3540°F (1950°C). When platinum thermocouples are used at temperatures exceeding 2200°F (1200°C), mullite should not be used, because it contains impurities which can contaminate platinum. For such applications, high-purity alumina is the proper choice.

Dual protection tubes may be used where the outer tube provides mechanical protection and the inner tube provides corrosion or permeation protection. See Figure 17c for an example of protective layers used on a high-temperature thermocouple.

The sheaths are either extruded or woven and are used to protect thermocouples or other wires. Woven sheaths are similar to those used for electrical insulation and may be made of stainless steel Inconel, tinned copper, fiberglass, or ceramics. They can withstand temperatures up to 2200°F (1204°C) continuously or 2800°F (1538°C) for short periods of time.

Extruded sheaths can be plastic or metallic. The metallic sheaths are commonly used for mineral-insulated thermocouples in which the wire elements are surrounded by ceramic oxides such as MgO (Figure 17d). After being packed with powdered oxides, the sheaths are swaged or rolled under pressure to reduce their diameter and tightly pack the powdered insulation. While this process serves to reduce intrusion of atmosphere within the sheath, it does not protect against humidity or moisture. Moisture intrusion is rapid and deteriorates insulation resistance to the point of thermocouple failure. Procedures for insulation resistance testing are to be found in ASTM documentation. Such testing is vital for mineral-insulated, metal-sheathed thermocouples stored in the open at construction sites or in repair shops for periods of

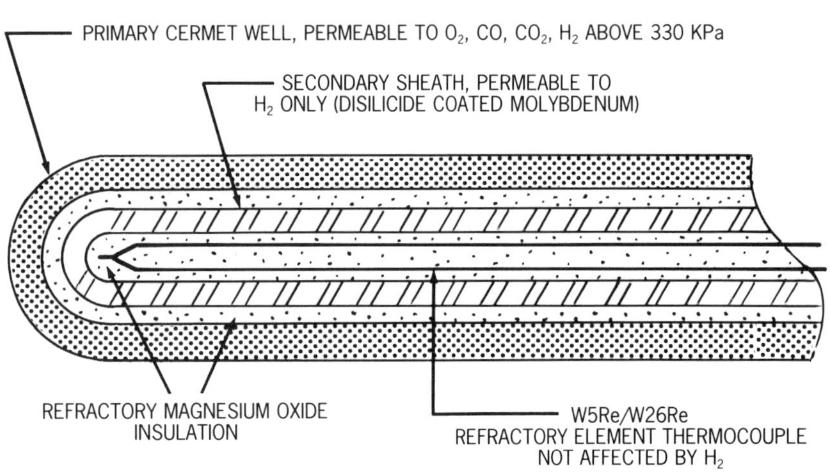

FIG. 17c
Multiple layers of protection used on a high-temperature thermocouple.

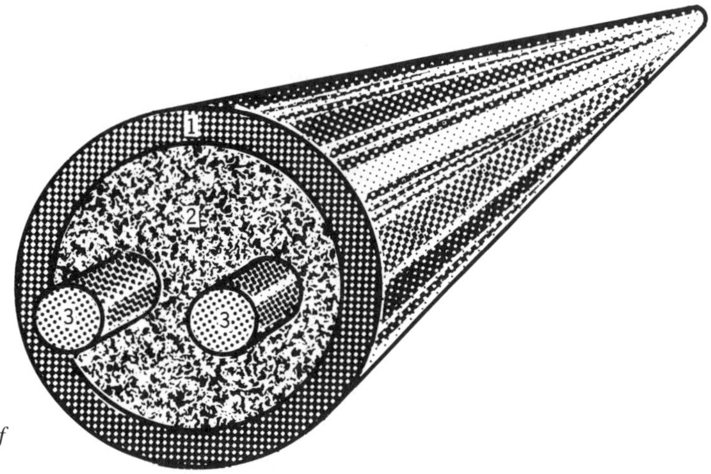

FIG. 17d
Thermocouple wires surrounded by magnesium oxide or other mineral insulation and protected by an integral metal sheath of stainless steel or Inconel.

time. Sheathing can also cause vapor transfer at elevated temperatures, which can result in thermocouple drift. This is the case with stainless steel sheaths used for nickel-bearing thermocouples such as types K and N. The offending element is manganese, and the problem can be eliminated by specifying low-manganese Inconel or modified Nicrosil sheathing.

THERMOWELL INSTALLATION

Well installation should be planned so that the wells are readily accessible for servicing and so that the thermal elements can be withdrawn without obstruction by adjacent structures. It is desirable to install spare wells for calibration purposes so the working sensor can be compared to a standard temperature sensor.

Vertical installation is preferred for very high temperature use to prevent sagging. Horizontal installation, on the other hand, avoids some contamination by foreign materials (Figure 17e). In horizontal installations, consideration should be given to the possibility that construction or repair personnel might misuse the wells as steps.

The thermowell immersion depth (U) should be sufficient to eliminate conduction error. A general rule is to use an

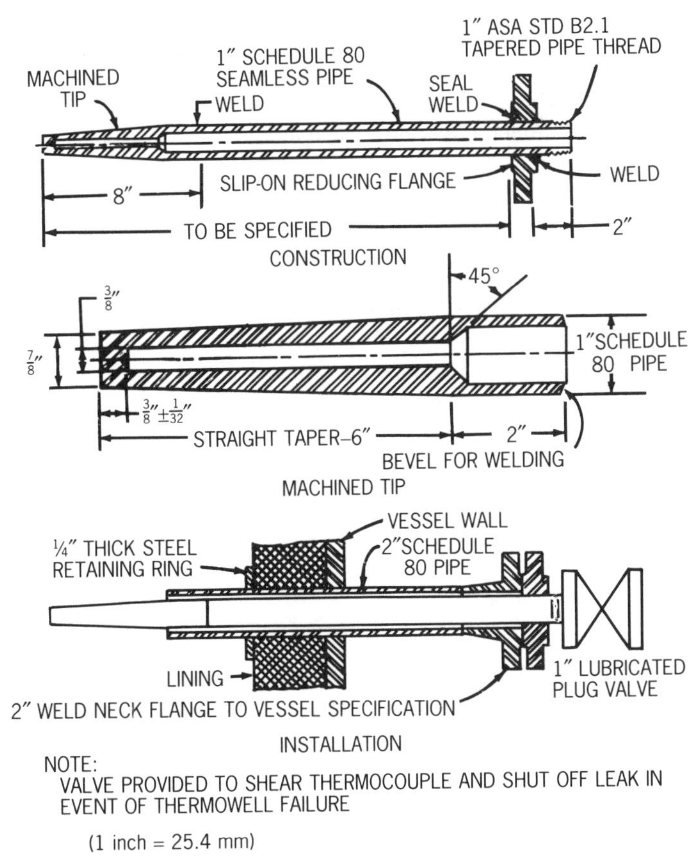

FIG. 17e
Flanged thermowell designs used on erosive services.[1]

insertion length equaling a minimum of 10 times the diameter of the protection tube or well. Another rule of thumb is to have the sensitive portion of the sensor immersed to a depth of a minimum of 3 in. (75 mm) plus the length of the sensitive portion. In the case of expansion bulbs, the immersion depth may be specified by the supplier or can be indicated on a calibration report, if furnished.

For pipe installations such as steam or hot water lines, insertion in an elbow on the axis of the pipe can permit an appropriate immersion depth should the diameter of the pipe be inadequate (Figure 17f).

When installing thermowells at an angle or in elbows it is hard to maintain sanitary conditions, which is an important consideration in the food and pharmaceutical industries. Thermowells inserted into small pipes can also cause excessive flow restrictions, pressure drops, or even plugging.

OTHER CONSIDERATIONS

The time constant of a thermal element increases with its mass. If time constants of a couple of seconds or less are required, unprotected, small diameter, bare thermal elements must be used. The addition of even a thin-walled sheath or thermowell increases the time constant to about 5 seconds as a minimum, and with larger bulbs it can reach 30 seconds. It is also important to maintain positive contact between the thermal element and the well in order to make sure that the temperature being measured reflects the temperature of the process on the outside and not the air temperature on the inside of the well.

If the process gases include hydrogen and a filled thermal element is used, it might be a good idea to vent the well, because hydrogen can damage the filling material.

For recommendations on well materials for a variety of services, refer to Table 17g.

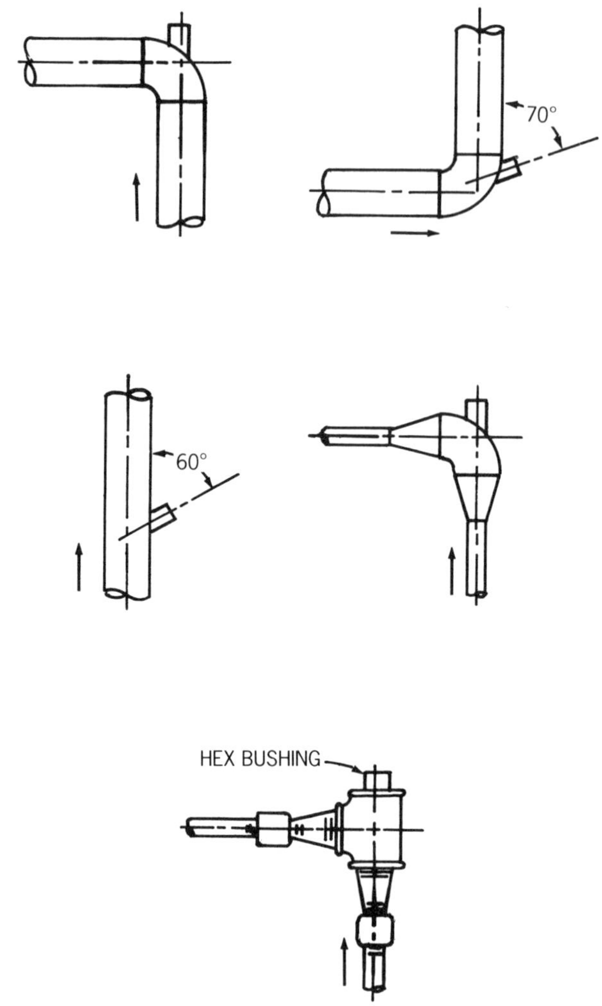

FIG. 17f
Alternate ways of installing thermowells in pipes that are 3 in. (75 mm) or smaller in diameter.[1]

TABLE 17g
Selection Guide for Thermowell Materials[2]

Application	*Protecting Tube Material*
Heat treating:	
Annealing	
Up to 704°C (1300°F)	black steel
Over 704°C (1300°F)	Inconel 600,[a] Type 446 SS
Carburizing hardening	
Up to 816°C (1500°F)	black steel, Type 446 SS
1093°C (1500 to 2000°F)	Inconel 600, Type 446 SS
Over 1093°C (2000°F)	ceramic[b]
Nitriding salt baths	Type 446 SS
Cyanide	nickel (CP)
Neutral	Type 446 SS
High speed	ceramic[b]

TABLE 17g Continued
Selection Guide for Thermowell Materials[2]

Application	Protecting Tube Material
Iron and steel:	
Basic oxygen furnace	quartz
Blast furnaces	
Downcomer	Inconel 600, Type 446 SS
Stove dome	silicon carbide
Hot blast main	Inconel 600
Stove trunk	Inconel 600
Stove outlet flue	black steel
Open hearth	
Flues and stack	Inconel 600, Type 446 SS
Checkers	Inconel 600, Cermets
Waste heat boiler	Inconel 600, Type 446 SS
Billet heating slab heating and butt welding	
Up to 1093°C (2000°F)	Inconel 600, Type 446 SS
Over 1093°C (2000°F)	silicon ceramic carbide[b]
Bright annealing batch	
Top work temperature	not required (use bare Type J thermocouple)
Bottom work temperature	Type 446 SS
Continuous furnace section	Inconel 600, ceramic[b]
Forging	silicon carbide, ceramic[b]
Soaking pits	
Up to 1093°C (2000°F)	Inconel 600
Over 1093°C (2000°F)	silicon ceramic carbide[b]
Nonferrous metals:	
Aluminum	
Melting	cast iron (white-washed)
Heat treating	black steel
Brass or bronze	not required (use dip-type thermocouple)
Lead	Type 446 SS, black steel
Magnesium	black steel, cast iron
Tin	extra heavy carbon steel
Zinc	extra heavy carbon steel
Pickling tanks	chemical lead
Cement:	
Exit flues	Inconel 600, Type 446 SS
Kilns, heating zone	Inconel 600
Ceramic:	
Kilns	ceramic[b] and silicon carbide[c]
Dryers	silicon carbide, black steel
Vitreous enameling	Inconel 600, Type 446 SS
Glass:	
Fore hearths and feeders	platinum thimble
Lehrs	black steel
Tanks	
Roof and wall	ceramic[a]
Flues and checkers	Inconel 600, Type 446 SS
Paper:	
Digesters	Type 316 SS, Type 446 SS
Petroleum:	
Dewaxing	Types 304, 310, 316, 321, 347 SS, carbon steel

TABLE 17g Continued

Selection Guide for Thermowell Materials[2]

Application	*Protecting Tube Material*
Petroleum cont.	
Towers	Types 304, 310, 316, 321, 347 SS, carbon steel
Transfer lines	Types 304, 310, 316, 321, 347 SS, carbon steel
Factioning column	Types 304, 310, 316, 321, 347 SS, carbon steel
Bridgewall	Types 304, 310, 316, 321, 347 SS, carbon steel
Power:	
Coal-air mixtures	304 SS
Flue gases	black steel, Type 446 SS
Preheaters	black steel, Type 446 SS
Steel lines	Types 347 or 316 SS
Water lines	low carbon steels
Boiler tubes	Types 304, 309, or 310 SS
Gas producers:	
Producer gas	Type 446 SS
Water gas	
Carburetor	Inconel 600, Type 446 SS
Superheater	Inconel 600, Type 446 SS
Tar stills	low carbon steels
Incinerators:	
Up to 1093°C (2000°F)	Inconel 600, Type 446 SS
Over 1093°C (2000°F)	ceramic (primary) silicon carbide (secondary)[a]
Food:	
Baking ovens	black steel
Charretort, sugar	black steel
Vegetables and fruit	Type 304 SS
Chemical:	
Acetic acid	
10 to 50% 21°C (70°F)	Type 304, Hastelloy C,[d] Monel
50% 100°C (212°F)	Type 316, Hastelloy C,[d] Monel
99% 21 to 100°C (70 to 212°F)	Type 430, Hastelloy C,[d] Monel
Alcohol, ethyl, methyl	
21 to 100°C (70 to 212°F)	Type 304
Ammonia	
All concentration 21°C (70°F)	Types 304, 316 SS
Ammonium chloride	
All concentration 100°C (212°F)	Type 316 SS, Monel
Ammonium nitrate	
All concentration 21 to 100°C (70 to 212°F)	Type 316 SS
Ammonium sulphate, 10% to saturated 100°C (212°F)	Type 316 SS
Barium chloride, all concentration, 21°C (70°F)	Monel, Hastelloy C
Barium hydroxide, all concentration, 21°C (70°F)	low carbon steels
Barium sulphite	Nichrome,[a] Hastelloy C
Brines	Monel
Bromine	tantalum, Monel
Butadiene	Type 304 SS
Butane	Type 304 SS

TABLE 17g Continued

Selection Guide for Thermowell Materials[2]

Application	*Protecting Tube Material*
Chemical cont.	
Butyl acetate	Monel
Butyl alcohol	copper, Type 304 SS
Calcium chlorate, dilute, 21 to 66°C (70 to 150°F)	Type 304 SS
Calcium hydroxide	
10 to 20% 100°C (212°F)	Type 304 SS, Hastelloy C
50% 100°C (212°F)	Type 316 SS, Hastelloy C
Carbolic acid, all, 100°C (212°F)	Type 316 SS
Carbon dioxide, wet or dry	2017-T4 aluminum, Monel, nickel
Chlorine gas	
Dry, 21°C (70°F)	Type 316 SS, Monel
Moist, 7 to 100°C (20 to 212°F)	Hastelloy C
Chromic acid, 10 to 50% 100°C (212°F)	Type 316 SS, Hastelloy C (all concentrations)
Citric acid	
15% 21°C (70°F)	Type 304 SS, Hastelloy C (all concentrations)
15% 100°C (212°F)	Type 316 SS, Hastelloy C (all concentrations)
Concentrated, 100°C (212°F)	Type 316 SS, Hastelloy C (all concentrations)
Copper nitrate	Types 304 SS, 316 SS
Copper sulphate	Types 304 SS, 316 SS
Cresols	Type 304 SS
Cyanogen gas	Type 304 SS
Dow therm[f]	low carbon steels
Ether	Type 304 SS
Ethyl acetate	Monel, Type 304 SS
Ethyl chloride, 21°C (70°F)	Type 304 SS, low carbon steel
Ethyl sulphate, 21°C (70°F)	Monel
Ferric chloride, 5% 21°C (70°F) to boiling	tantalum, Hastelloy C
Ferric sulphate, 5% 21°C (70°F)	Type 304 SS
Ferrous sulphate, dilute, 21°C (70°F)	Type 304 SS
Formaldehyde	Types 304 SS, 316 SS
Formic acid, 5% 21 to 66°C (70 to 150°F)	Type 316 SS
Freon	Monel
Gallic acid, 5% 21 to 66°C (70 to 150°F)	Monel
Gasoline, 21°C (70°F)	Type 304 SS, low carbon steel
Glucose, 21°C (70°F)	Type 304 SS
Glycerine, 21°C (70°F)	Type 304 SS
Glycerol	Type 304 SS
Hydrobromic acid, 98% 100°C (212°F)	Hastelloy B
Hydrochloric acid	
1%, 5% 21°C (70°F)	Hastelloy C
1%, 5% 100°C (212°F)	Hastelloy B
25% 21 to 100°C (70 to 212°F)	Hastelloy B
Hydrofluoric acid, 60% 100°C (212°F)	Hastelloy C, Monel
Hydrogen peroxide, 21 to 100°C (70 to 212°F)	Types 316 SS, 304 SS
Hydrogen sulphide, wet and dry	Type 316 SS
Iodine, 21°C (70°F)	tantalum
Lactic acid	
5% 21°C (70°F)	Type 304 SS, 316 SS
5% 66°C (150°F)	Type 316 SS
10% 100°C (212°F)	tantalum
Magnesium chloride	
5% 21°C (70°F)	Monel, nickel
5% 100°C (212°F)	nickel
Magnesium sulphate, hot and cold	Monel

TABLE 17g Continued

Selection Guide for Thermowell Materials[2]

Application	*Protecting Tube Material*
Chemical cont.	
Muriatic acid, 21°C (70°F)	tantalum
Naptha, 21°C (70°F)	Type 304 SS
Natural gas, 21°C (70°F)	Types 304 SS, 316 SS, 317 SS
Nickel chloride, 21°C (70°F)	Type 304 SS
Nickel sulphate, hot and cold	Type 304 SS
Nitric acid	
5% 21°C (70°F)	Types 304 SS, 316 SS
20% 21°C (70°F)	Types 304 SS, 316 SS
50% 100°C (70°F)	Types 304 SS, 316 SS
50% 100°C (212°F)	Types 304 SS, 316 SS
65% 100°C (212°F)	Type 316 SS
Concentrated, 21°C (70°F)	Types 304 SS, 316 SS
Concentrated, 100°C (212°F)	tantalum
Nitrobenzene, 21°C (70°F)	Type 304 SS
Oleic acid, 21°C (70°F)	Type 316 SS
Oleum, 21°C (70°F)	Type 316 SS
Oxalic acid	
5% hot and cold	Type 304 SS
10% 100°C (212°F)	Monel
Oxygen	
21°C (70°F)	steel
Liquid	SS
Elevated temperatures	SS
Palmitic acid	Type 316 SS
Pentane	Type 340 SS
Phenol	Types 304 SS, 316 SS
Phosphoric acid	
1%, 5% 21°C (70°F)	Type 304 SS
10% 21°C (70°F)	Type 316 SS
10% 100°C (212°F)	Hastelloy C
30% 21 to 100°C (70 to 212°F)	Hastelloy B
85% 21 to 100°C (70 to 212°F)	Hastelloy B
Picric acid, 21°C (70°F)	Type 304 SS
Potassium bromide, 21°C (70°F)	Type 316 SS
Potassium carbonate, 1% 21°C (70°F)	Types 304 SS, 316 SS
Potassium chlorate, 21°C (70°F)	Type 304 SS
Potassium hydroxide	
5% 21°C (70°F)	Type 304 SS
25% 100°C (212°F)	Type 304 SS
60% 100°C (212°F)	Type 316 SS
Potassium nitrate	
5% 21°C (70°F)	Type 304 SS
5% 100°C (212°F)	Type 304 SS
Potassium permanganate, 5% 21°C (70°F)	Type 304 SS
Potassium sulphate, 5% 21°C (70°F)	Types 304 SS, 316 SS
Potassium sulphide, 21°C (70°F)	Types 304 SS, 316 SS
Propane	Type 304 SS, low carbon steel
Pyrogallic acid	Type 304 SS
Quinine bisulphate, dry	Type 316 SS
Quinine sulphate, dry	Type 304 SS
Seawater	Monel or Hastelloy C
Salicylic acid	nickel
Sodium bicarbonate	
All concentration, 21°C (70°F)	Type 304 SS
5% 66°C (150°F)	Types 304 SS, 316 SS

TABLE 17g Continued
Selection Guide for Thermowell Materials[2]

Application	Protecting Tube Material
Chemical cont.	
Sodium carbonate, 5% 21 to 66°C (70 to 150°F)	Types 304 SS, 316 SS
Sodium chloride	
5% 21 to 66°C (70 to 150°F)	Type 316 SS
Saturated, 21 to 100°C (70 to 212°F)	Type 316 SS, Monel
Sodium fluoride, 5% 21°C (70°F)	Monel
Sodium hydroxide	Types 304 SS, 316 SS, Hastelloy C
Sodium hypochlorite, 5% still	Type 316 SS, Hastelloy C
Sodium nitrate, fused	Type 316 SS
Sodium peroxide	Type 304 SS
Sodium sulphate, 21°C (70°F)	Types 304 SS, 316 SS
Sodium sulphide, 21°C (70°F)	Type 316 SS
Sodium sulphite, 30% 66°C (150°F)	Type 304 SS
Sulphur dioxide	
Moist gas, 21°C (70°F)	Type 316 SS
Gas, 302°C (575°F)	Types 304 SS, 316 SS
Sulphur	
Dry molten	Type 304 SS
Wet	Type 316 SS
Sulphuric acid	
5% 21 to 100°C (70 to 212°F)	Hastelloy B, 316 SS
10% 21 to 100°C (70 to 212°F)	Hastelloy B
50% 21 to 100°C (70 to 212°F)	Hastelloy B
90% 21°C (70°F)	Hastelloy B
90% 100°C (212°F)	Hastelloy D
Tannic acid, 21°C (70°F)	Type 304 SS, Hastelloy B
Tartaric acid	
21°C (70°F)	Type 304 SS
66°C (150°F)	Type 316 SS
Toluene	2017-T4 aluminum, low carbon steel
Turpentine	Types 304 SS, 316 SS
Whiskey and wine	Type 304 SS, nickel
Xylene	copper
Zinc chloride	Monel
Zinc sulphate	
5% 21°C (70°F)	Types 304 SS, 316 SS
Saturated, 21°C (70°F)	Types 304 SS, 316 SS
25% 100°C (212°F)	Types 304 SS, 316 SS

[a]Trademark of the International Nickel Co.
[b]Due to susceptibility to cracking, sudden thermal shocks should be avoided.
[c]Due to susceptibility to cracking, sudden thermal shocks should be avoided.
[d]Trademark of the Cabot Corp.
[e]Trademark of the Driver-Harris Co.
[f]Trademark of the Dow Chemical Corp.

References

1. American Petroleum Institute, "Manual on Installation of Refinery Instruments," API RP 550, March 1965.
2. ASTM, "Manual on the Use of Thermocouples in Temperature Measurement," STP 470B.

Bibliography

ASME Boiler and Pressure Vessel Code, Section VIII, 1965.

ASTM Standards, Part 44.

Daneman, H.L., "MIMS Thermocouple Stability," *Measurements and Control,* February 1991.

Gordon Co., *Temperature Measurement Handbook.*

JMS Southeast Co. Catalog.

Masek, J.A., "Thermowell Design for Process Piping," *Hydrocarbon Processing,* February 1964.

Nicholas, J.V., and White, D.R., "Traceable Temperatures," 1982.

OMEGA Engineering Inc., *Complete Temperature Measurement Handbook and Encyclopedia,* 1989.

Thermo Electric Co., "Temperature Measurement Designers' Guide."

18 Ultrasonic Thermometers

H. L. DANEMAN (1993)

Flow Sheet Symbol

Applications:	Temperature measurement within solids, inside nuclear reactors, and in furnaces
Types:	Piezoelectric and magnetostrictive
Temperature Ranges:	From cryogenic to 20,000°C (36,000°F)
Cost:	\$10,000 or more
Partial List of Suppliers:	Panametrics Inc.; Scientific Engineering Instruments

The relationship between sound velocity and temperature has been known for nearly 100 years, but until recently it has been little used. Temperature dependence of velocity in an ideal gas is expressed as

$$v^2 = \frac{\gamma RT}{M_w} \qquad \mathbf{18(1)}$$

where v is sound velocity, γ is the ratio of specific heats, R is the gas constant per mole, M_w is the molecular weight, and T is the absolute temperature.

A method of temperature measurement in a plasma jet involves the use of two quartz probes set a fixed distance apart. The sound velocity is determined by circuitry for the continuous measurement of the ultrasonic wave transit time (see Figure 18a).

Making the measurement one of time is an advantage resulting in high accuracy. Because the gas is the thermometer element, errors such as leakage are absent and fast changes can be followed. Disadvantages include high cost, non-ideal gas behavior, pressure correction, accurate determination of γ, and the inability to make point measurements.

Acoustical temperature sensors can measure temperature from the cryogenic range to plasma levels (20,000°C, or 36,000°F). The accuracy can approach that of a primary standard as in the example of the ultrasonic interferometer developed by Harmon Plumb and George Cataland of the NIST. A fluidic oscillator in which the resonant frequency of the cavity varies with temperature can be employed as a high-temperature pyrometer.

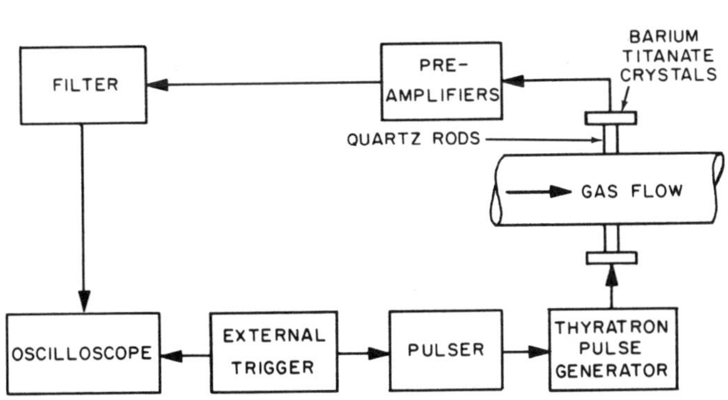

FIG. 18a
Simplified ultrasonic temperature measuring apparatus.

Ultrasonic thermometers are based on the effect of the temperature on the velocity of sound waves in the medium transmitting the sound waves. Temperature measurements can be made in gases, liquids, or solids. The relationship between sound velocity and temperature for various materials is shown in Figure 18b.

The acoustic velocity can be detected by immersing a rod or wire into the fluid or by using the medium itself as an acoustic conductor. The sensor rod can measure the temperature at a point or, by means of a series of constrictions or indents, can profile or average the temperature within the medium. This is done by measuring the time lags of the sound waves reflected from the consecutive indents (see Figure 18c).

When the medium is used as the conductor of the ultrasonic pulse, the transducer is located within the vessel or on its external shell. This latter configuration is useful when measuring the temperature of solids or extremely hot or corrosive materials such as molten sodium.

The acoustic pulse is generated by a piezoelectric crystal cut to resonate at a frequency ranging from 0.5 to 3 MHz or by means of magnetostrictive materials.

The thin wire sensor (Figure 18c) is installed like a thermocouple. A lead-in wire carries the pulse to the thin wire made of a material suitable for the process medium and its temperature range. Reflection from the beginning and the end of the thin wire provides the time lapse information for temperature determination. The selection of materials is more flexible than for thermocouples as only one material is involved.

Noninvasive applications employ a transmitting trans-

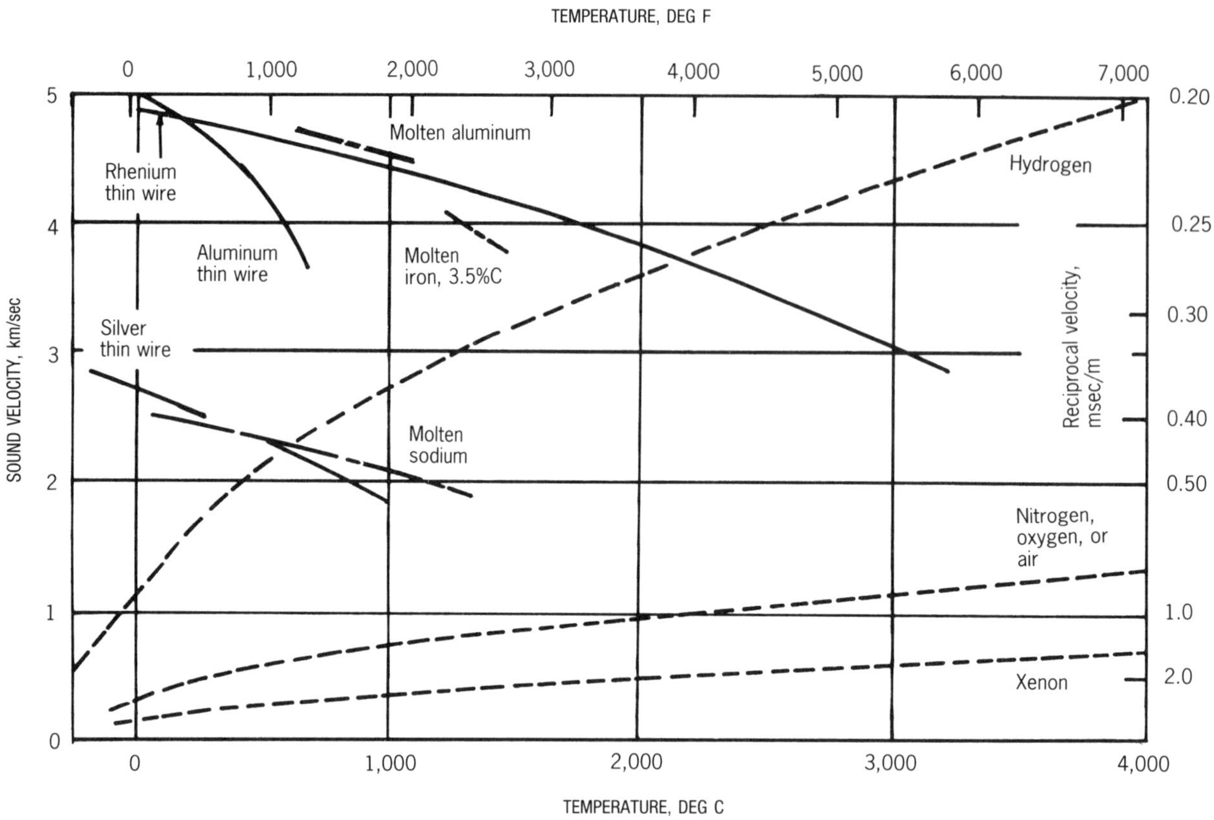

FIG. 18b
The velocity of sound increases in gases and decreases in liquids and solids as the temperature rises.[1]

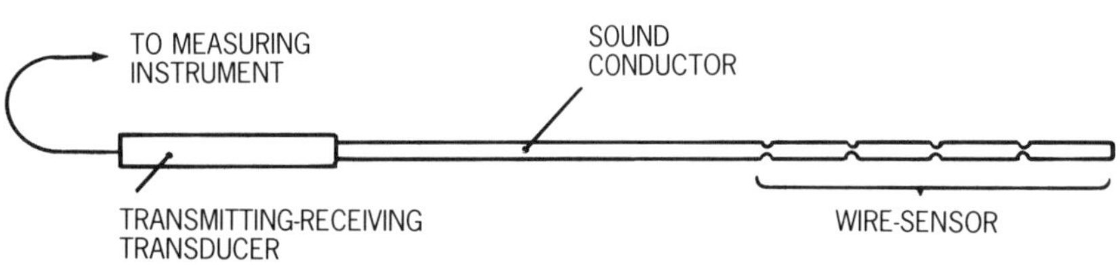

FIG. 18c
Ultrasonic thermometer probe.[2]

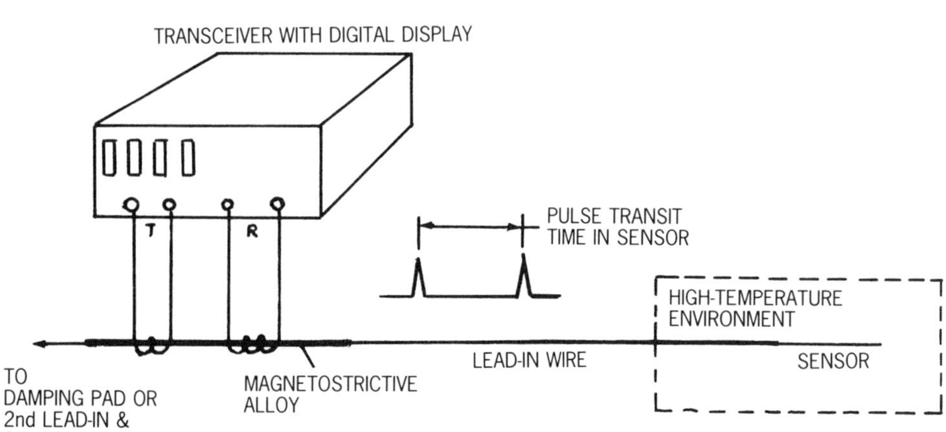

FIG. 18d
Thin wire ultrasonic thermometer with magnetostrictive transducer. (Courtesy of Panametrics)

TABLE 18e
*Applications for Ultrasonic Thermometers**

Measured Object	*Temperature (°C)*	*Sensor or Medium*	*Sound Conductor*	*Transducer*	*Signal*
Gas	1500–15,000	N_2	Quartz	Piezoelectric	One period 1 MHz
Liquid	1000	Molten Na	Stainless steel	Piezoelectric	One period or period series 3–10 MHz
Solids with holes	3000	Re-wire	Tungsten	Magnetostrictive	Period series 0.1 MHz
Solids without holes	1500	Steel wire	Steel	Piezoelectric	Period series 1 MHz

*From Reference 3.

ducer to send acoustic energy through the medium. Receiving transducers detect the energy and the time delay as measured by an intervalometer, which determines the velocity of sound and therefore the temperature of the process. A thermometer employing the magnetostrictive transducer is shown in Figure 18d.

Ultrasonic thermometry can be used at temperature extremes, in high electrical fields, or when the medium being measured is inaccessible. It is also useful for averaging the temperature of bulk materials or for profiling furnace temperatures. Other applications are tabulated in Table 18e.

References

1. Lynnworth, L.C., ''High Temperature Acoustics,'' Naval Research Reviews, 1970.
2. Michalski, L., Eckersdorf, K., and McGhee, J., ''Temperature Measurement,'' New York: John Wiley & Sons, 1991.
3. Lynnworth, L.C., and Carnavale, E.H., ''Ultrasonic Thermometry Using Pulse Techniques,'' Temperature 1972.

Bibliography

Lynnworth, L.C., ''Temperature Profiling Using Multizone Ultrasonic Waveguides,'' *Temperature,* 1982.

Colclough, A.R., ''Primary Acoustic Thermometry: Principles and Current Trends,'' *Temperature,* 1982.

Lynnworth, L.C., ''Ultrasonic Measurements for Process Control,'' Academic Press, 1989.

Lynnworth, L.C., Papadakis, E.P., Patch, D.R., Fowler, K.A., and Shepard, R.L., ''Nuclear Reactor Applications of New Ultrasonic Transducers,'' IEEE Transactions on Nuclear Science, 1970.

INDEX